# Neues verkehrswissenschaftliches Journal

**Ausgabe 18**

# Metaheuristic-based Dispatching Optimization Integrated in Multi-scale Simulation Model of Railway Operation

Von der Fakultät Bau- und Umweltingenieurwissenschaften

der Universität Stuttgart zur Erlangung der Würde eines

Doktor-Ingenieurs (Dr.-Ing.) genehmigte Abhandlung

Vorgelegt von

## Jiajian Liang

aus Baicheng, VR China

Hauptberichter:     Prof. Dr.-Ing. Ullrich Martin

Mitberichter:     Prof. Dr.-Ing. Thomas Siefer

Tag der mündlichen Prüfung:     23.02.2017

M.Sc. Jiajian Liang

Institut für Eisenbahn- und Verkehrswesen der Universität Stuttgart

**February 2017**

# Preface

Although operation in the field of public railways is highly determinant, it is fundamentally a strongly stochastically-influenced process due to diverse external and internal influences. Accordingly, its rescheduling should punctually recognize potential deviations during operation and initiate targeted measures in order to avoid the deviations from becoming effective, or at least to minimize their impact.

Despite the fact that there are currently computerized tools available for dispatching, the actual dispatching process is still directly dependent upon the operator. The increasingly centralized operational control systems and tendentially-growing operating areas exacerbate timely and sound dispatching reactions on the part of the operator when several or complex conflicts, which have wide-reaching repercussions, occur. Furthermore an operator-independent, automated dispatching algorithm is an urgent requirement for an adequate and realistic simulation of railway operation.

The since-concluded project of the DFG (MA 2326/15-1) „Der Einfluss der Disposition auf den Zusammenhang zwischen Belastung und Betriebsqualität von Eisenbahnsystemen" is focused predominantly towards a condition-oriented solution of conflicts through the use of a greedy algorithm. This approach was further developed, and its applicability expanded, in the dissertation from Mr. Liang through the use of metaheuristic methodology.

Through the findings elaborated in the dissertation, for the methodological expansion of the methodology of multi-scale simulation and dispatching in rail services through the situation-dependent, local, flexibly-bounded, microscopic observation under consideration of various signaling systems, a significant contribution was acquired in the process of the automated management and enhancement of efficiency of software solutions for railway operation simulation with immediate, practically-oriented, relevance. The particularly commendable innovations are the improvement of those parameters that were known, but that had not yet been implemented in the form of an algorithm with dispatching, as well as the improvement of the validity of the proposals made by the developed dispatching algorithm for the resolution of conflicts during ongoing railway operations. Through the use of the developed algorithms, the multi-scalability that was previously only considered for auxiliary structures and was unpresentable, may, in the future, be directly integrated into the rescheduling procedure as well as in the simulation process.


Stuttgart, April 2017

Ullrich Martin

## Acknowledgement

First of all, I would like to express my gratitude to Prof. Ulrich Martin, my doctoral supervisor. He provided me the opportunity to apply for the top research fund (DFG) in Germany. This allowed me to gain valuable experience in applying for research projects, and also gave me the chance to become a member of the research staff of our institute. In the final stage of my dissertation, Prof. Martin gave me lots of timely encouragement, which enabled me to complete the dissertation on time and to successfully pass the doctoral exam. Without the help of Prof. Martin, it is hard to imagine how I could have accomplished all these tasks. At the same time, I am also very grateful for the support of Prof. Siefer, my second supervisor.

Secondly, I would like to offer a special thanks to my colleague, Yong Cui. He gave me the framework of the simulation model developed in his own dissertation; hence I could quickly start developing the multi-scale simulation model for my dissertation. In the whole process of algorithm design and model development, his help has been irreplaceable.

Thirdly, my thanks would go to my former colleague, Zifu Chu. At the beginning of my doctoral study, he patiently explained the codes to me; this made me quickly familiar with a new programming language. For algorithm design, he was also of great help. When I was in charge of organizing lectures for our institute, my colleague, Carlo von Molo also provided me a lot of help. His support allowed me to overcome difficulties and successfully complete the tasks required.

Finally, I sincerely thank my wife Pengfei Wang, who has accompanied me throughout these arduous but meaningful years.

# Table of Content

# List of Figures

# List of Tables

# Kurzfassung

Die Dispositionssysteme dienen als unmittelbarer Bestandteil des Bahnbetriebs der Verminderung negativer Auswirkungen unvorhergesehener Ereignisse auf den Betriebsprozess. Aufgrund der zeitkritischen Entscheidungsfindung und der damit verbundenen Komplexität der prozessintegrierten Disposition muss regelmäßig ein tragfähiger Kompromiss zwischen der oftmals durch die Rechentechnik determinierten Bearbeitungszeit einer Dispositionsaufgabe und der Qualität des Dispositionsergebnisses gefunden werden. Dies trifft gleichermaßen für Software zur Simulation des Bahnbetriebes zu, um den Arbeitsaufwand betrieblicher Untersuchungen vertretbar zu gestalten. Dementsprechend ist es bei der Gestaltung von Disposition-Tools besonders wichtig, einen ausgewogenen Ausgleich zwischen benötigter Rechenzeit und hinreichender Qualität des Ergebnisses zu finden. Mit dieser zentralen Zielstellung wurde in dieser Dissertation ein Dispositionsoptimierungsalgorithmus entwickelt, welcher auf einem weit verbreiteten metaheuristischen Algorithmus, der Tabu-Suche, und der Integration in ein mehrstufiges Simulationsmodell basiert. Der verfolgte Ansatz basiert unmittelbar auf Erkenntnissen aus dem DFG-Projekt "Der Einfluss der Disposition auf den Zusammenhang zwischen Belastung und Betriebsqualität von Eisenbahnsystemen" und erweitert diese um ein universelles Mehrskalenmodell [Martin und Liang, 2014].

Das verwendete Mehrskalen-Simulationsmodell zeichnet sich durch eine kontinuierliche Skalierung aus, bei der Bahnbetriebsprozesse gleichzeitig auf mikroskopischer, mesoskopischer und makroskopischer Ebene simuliert werden. Für große Untersuchungsräume werden relevante Bereiche auf mikroskopischer Ebene betrachtet und die anderen Bereiche effizient auf mesoskopischen und makroskopischen Ebenen dargestellt. Darüber hinaus wurde eine Bewertungsmethode für das Mehrskalenmodell entwickelt, um signifikante Werte verschiedener Infrastrukturelemente im Untersuchungsraum zu ermitteln. In Abhängigkeit von diesen signifikanten Werten kann das Simulationsmodell kontinuierlich zwischen drei Abstraktionsebenen migrieren, so dass die rechnerische Komplexität und die Genauigkeit der Simulationsergebnisse in diesem Modell gut ausgeglichen sind.

Mit dem vorgeschlagenen Mehrskalen-Simulationsmodell kann die Reihenfolge der Zugbewegungen durch das einfachste Dispositionsprinzip (First Come First Serve) oder eine vordefinierte Dispositionslösung bestimmt werden. „First Come First Serve" wird eingesetzt, um grundlegende Dispositionslösungen zu generieren, während vordefinierte Dispositionslösungen genutzt werden, um optimierte Lösungen zu simulieren und zu bewerten. Der vom Simulationsmodell unterstützte, auf der Tabu-Suche basierende Algorithmus zur Dispositionsoptimierung, ist in der Lage, die Basislösung durch eine Reihe von Dispositionsmaßnahmen iterativ zu optimieren, bis eine zufriedenstellende Lösung erreicht wird. Es konnte an einem Referenzbeispiel nachgewiesen werden, dass der entwickelte Algorithmus zur Dispositionsoptimierung eine suboptimale/optimale Lösung in einer begrenzten Zeit bereitstellen kann.

# Abstract

The dispatching system serves as an integral component of railway operation control and aims to eliminate the negative impacts of unforeseen events occurred during the operation process. On account of the time-critical decision-making and the associated complexity of the process-integrated dispatching, an acceptable compromise must be found regularly between the processing time of a dispatching task, which is often determined by the computer technology, and the dispatching solution quality. This applies equally to simulation software of railway operation, in order to make the workload of operational investigations acceptable. Accordingly, it is particularly important in the design of dispatching tools to find a good balance between required computation time and sufficient quality of results. With this central goal, a dispatching optimization algorithm was developed in this dissertation, which is based on a widely used metaheuristic algorithm – tabu search – and the integration in a multi-scale simulation model. The approach is based directly on the findings from the DFG project "The influence of dispatching on the relationship between capacity and operation quality of railway systems" and expands these by a universal multi-scale model [Martin and Liang, 2017].

The multi-scale simulation model is characterized by continuously scaling, in which railway operation processes are simulated on microscopic, mesoscopic and macroscopic levels concurrently. For large investigation areas, the relevant areas are presented on microscopic level, while the others are presented on more efficient mesoscopic and macroscopic levels. Furthermore, an assessment method for the multi-scale model was developed to determine the significant values of different infrastructure elements in the investigation area. Depending on the significant values, the simulation model can migrate continuously between three abstraction levels, so that the computational complexity and the accuracy of simulation results are well-balanced.

With the proposed multi-scale simulation model, the sequence of train movements can be determined by the simplest dispatching principle (First Come First Serve) or a predefined dispatching solution. "First Come First Serve" is employed to generate basic dispatching solutions, while predefined dispatching solutions are used to simulate and evaluate optimized solutions. The simulation model-supported tabu search-based algorithm for dispatching optimization is able to optimize the basic solution by a series of dispatching measures iteratively until a satisfactory solution is obtained. It could be proved by means of a reference example that the developed algorithm for dispatching optimization can provide a suboptimal/optimal solution in a limited time.

# 1 Introduction

A railway system is a complex system consisting of infrastructure, vehicle and operation components. With the assistance of a railway operation control system, railway traffic is managed according to an operation plan (usually named as schedule). In order to optimize the use of infrastructure network capacity and ensure railway service quality, the operation plan has to be carefully prepared in advance. During the operation process, especially in railway networks with complex topology and high traffic flow, disturbances are likely to occur, which may result in severe deviations of train movements from the pre-designed operation plan. Once conflicts have occurred or potential conflicts between trains have been detected, suitable dispatching actions should be executed to minimize the negative impacts of the disturbances.

Due to the inherent complexity of dispatching tasks, railway dispatching is nowadays still carried out in large part manually in real time, and the quality of the dispatching decisions highly depends on the experience of the dispatchers. In order to support dispatchers, some dispatching assistant tools have been developed or are being developed; such as a computer supported dispatching assistant KE/KL+ZLR, which was developed within the project Regler [Molo, 2017]. Dispatching tasks cover a wide field, such as train path and train priority sequence rescheduling, crew rescheduling and rolling stock circulation [Corman and Meng, 2013]. Based on the prior art the existing research has mainly attempted to solve a certain aspect of the problem, and a fully-automatic dispatching system is not yet realistic. The dispatching model developed in this dissertation aims to help dispatchers to optimize train paths and train priority sequences in case of disturbances during the operation process.

In the development of dispatching models, two aspects need be considered: modelling of railway operation and the dispatching optimization algorithm. The former is indispensable to accurately assess the impact of disturbances, and the latter is employed to find solutions with minimal impact. Based on the previous research of the IEV (Institute of Railway and Transportation Engineering at Universität Stuttgart, in German: *Institut für Eisenbahn- und Verkehrswesen der Universität Stuttgart*), synchronous simulation is employed to model railway operations. Simulation models can be classified into three types according to their levels of details of description: microscale, mesoscale and macroscale models. In order to balance accuracy and computation complexity a multi-scale simulation model was developed in this dissertation. This type of model was firstly proposed in [Cui and Martin, 2011], in which simulation is concurrently carried out on microscopic, mesoscopic and macroscopic levels. The multi-scale model is applied to both generations of basic dispatching solutions and evaluation of optimized dispatching solutions. Dispatching optimization is a typical combinatorial optimization problem, and exhaustive search becomes impractical when there is a large set of possible dispatching solutions. To speed up the search, a widely used metaheuristic algorithm – tabu search – was adopted as the basis of the dispatching optimization algorithm. In this algorithm, the optimization of train paths and train priority sequences are solved as a whole.

This dissertation addressed multi-scale simulation and dispatching optimization in railway operation, which shared the same fundamentals as the DFG project [Martin and Liang, 2017]. In the DFG project, a widely used heuristic algorithm – greedy algorithm – was employed as the basis of the dispatching optimization algorithm. Based on new findings from this project, the dispatching optimization algorithm was accordingly modified and expanded, and a more powerful metaheuristic algorithm – tabu search – was chosen as the basis for the algorithm in this dissertation. Taking the advantage of the special memory structures of tabu search, the search scope of the optimization algorithm is broadened and the problem of trapping in local optimal is solved. Furthermore, within the framework of the DFG project, a method of system state classification was developed, and the influences of dispatching on capacity and operation quality were systematically evaluated. However, these two topics will not be covered in this dissertation. For more details it is referred to [Martin and Liang, 2017]. The structure of this dissertation is organized as follows:

- Chapter 2 provides an overview of the railway operation control systems and the dispatching process in reality, along with an overview of railway operation modelling methods and dispatching optimization techniques in the existing researches.
- The components and workflow of a synchronous simulation model on different description levels (microscopic, mesoscopic and macroscopic level) is elaborated in Chapter 3. In this model, train movements are implicitly regulated by the simplest dispatching principle – First Come First Serve.

– An assessment method for the multi-scale model is developed in Chapter 4, which determines the significant value of each area in the entire investigated area, and accordingly decides the proper description level of each area.

– The multi-scale simulation model developed in Chapter 3 is modified and expanded in Chapter 5, in order to integrate the function of priority sequence control into the model. In the multi-scale simulation model with priority sequence control train movements are explicitly regulated by a pre-given dispatched timetable.

– In Chapter 6, a tabu search based dispatching optimization model is developed, and its performance is analyzed based on a series of test cases of a reference example. The main achievements in this dissertation and potential topics for further research are summarized in Chapter 7.

# 2 Basics of Railway Operation Control and Dispatching

The railway operation control system serves as a fundamental part of a railway system, and it guides train movements on infrastructures in a safe and efficient manner to realize the operation plan. To eliminate the unforeseen disturbances during the operation process, the dispatching module is indispensable as an essential component element of the railway operation control system. In order to design an automatic dispatching optimization system, it is necessary to have a basic understanding of railway operation control and dispatching. So in this chapter, an overview of the railway operation control system and the dispatching process will be provided in Section 2.1 and 2.2, and an overview on modelling methods of railway operation and dispatching optimization techniques will be given in Section 2.3 and Section 2.4, respectively.

## 2.1 Railway Operation Control

Railway traffic control is an important module of railway operation control. On the signal-controlled railway lines in Germany two types of traffic control authorities are in service at present: traffic control with local operators (in German: *Fahrdienstleiter*) and centralized traffic control (CTC).

### 2.1.1 Conventional Lines with Local Operators

On conventional signal-controlled railway lines, railway traffic control is realized with the assistance of local interlocking towers nearby tracks (Figure 2-1). The interlocking machines located in interlocking towers operate all points and signals in their respective operational territories, and interlocking towers are staffed with local operators. There are two types of local operators (train director and leverman) and two types of interlocking towers (command tower and dependent tower) at the German system. Command interlocking towers are staffed with train directors (e.g. Interlocking Tower 1 in Figure 2-1) and dependent towers with levermen (e.g. Interlocking Tower 2 in Figure 2-1). The train director is in full charge of issuing train movement authorities and communicating with dispatchers (i.e. running messages) and neighboring train directors (i.e. train messages). A leverman's duties are to set points and signals for train movements in accordance with the command of train directors, and to be responsible for authorizing shunting movements. On lines with high traffic flow, local operators are coordinated through dispatchers to avoid delays and congestions (e.g. Interlocking Tower 1, 2 and 3), while on lines with low traffic flow, the traffic is controlled without dispatchers (e.g. Interlocking Tower 4).

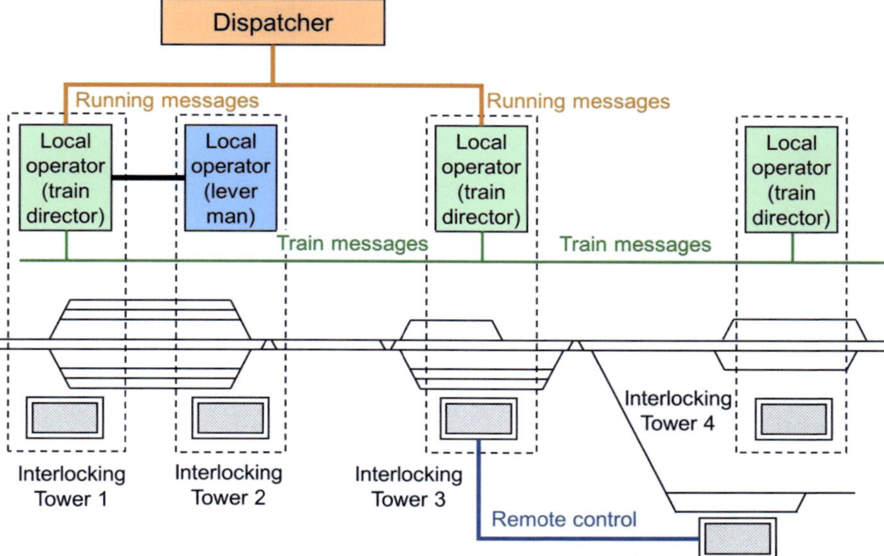


**Figure 2-1: Traffic Control on Conventional Signal-controlled Lines (modified from [PT1, 2016])**

## 2.1.2  CTC Lines

For CTC systems, railway lines are equipped with an electronic interlocking system, which enable the operation control to be executed remotely by operators working in control centers (Figure 2-2). Theoretically local operators (levermen) are only necessary for shunting tasks in large stations or junctions with complex topology and high traffic flow (e.g. the station on the left side in Figure 2-2). Furthermore, the conventional railway traffic control system can be well integrated with the CTC system (e.g. the station on the right side in Figure 2-2).

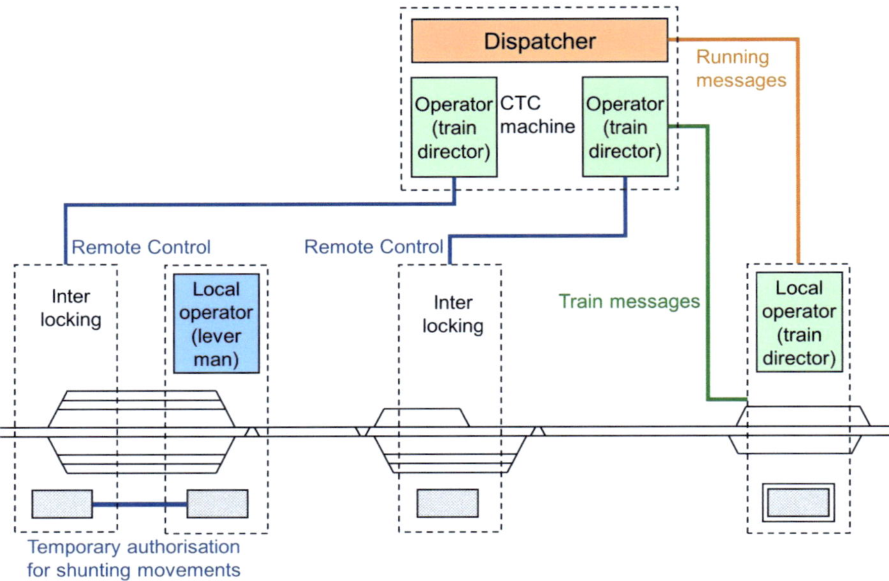



**Figure 2-2: Centralized Train Control (modified from [PT1, 2016])**

More specifically, the traffic regulation in Germany consists of two levels as shown in Figure 2-3 [DB NETZ AG, 420.02]: the network operation control center (in German: *Netzleitzentrale*) and seven regional operation control centers (in German: *Betriebszentrale*). The former is staffed with network coordinators (in German: *Netzkoordinator*), whose duty is to check and coordinate the necessary railway operations at the supra-regional and cross-border levels. The latter is staffed with regional dispatchers (in German: *Bereichsdisponenten*), traffic controllers (in German: *Zugdisponent* or *Zuglenker*) and operators (in Germany, *Fahrdienstleiter/ örtlich zuständiger Fdl*). Regional dispatchers are responsible for monitoring all train movements in the entire region, while traffic controllers are responsible for solving train path conflicts in the subordinate regulating areas. Operators are full in charge of authorizing both train and shunting movements [Pachl, 2002].

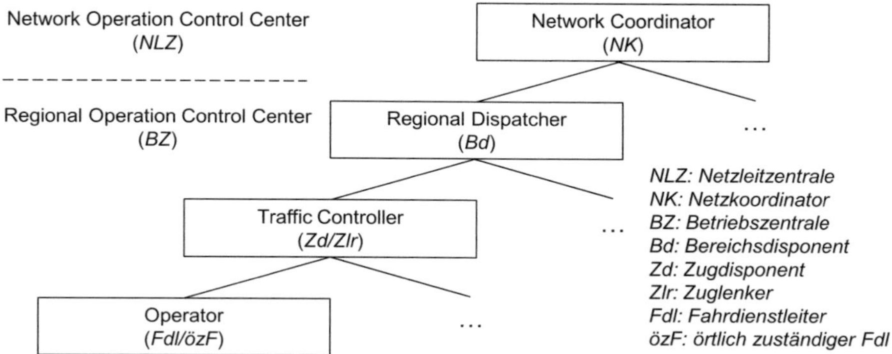


**Figure 2-3: The Structure of German Railway Traffic Regulation**

For both traffic control systems, especially CTC characterized by a large control territory, the implementation of a computer-supported dispatching optimization tool can, on one hand, improve the operation quality and, on the other hand, increase the infrastructure exploration rate and avoid redundant infrastructure investments.

## 2.2 General Process and Basic Methods of Railway Dispatching

Computer-based dispatching systems are used to assist dispatchers to identify and solve conflicts during the operation process. In general, the dispatching process can be summarized into five steps as shown in Figure 2-4 (similar descriptions of dispatching process can be found in [Cui, 2010], [D'Ariano, 2008] and [Lüthi, et al., 2007]). The train describer system equipped on conventional lines with local operators and CTC lines can support dispatchers in monitoring the current traffic situation. A train describer system can identify the current location of trains and occupation information of block sections, and show the information on the display on the panel of the operator or dispatcher [Pachl, 2002]. Based on the information provided by the train describer system, the current and the forecasted traffic situations can be displayed in the form of a traffic diagram in order to assist the dispatcher in identifying and solving conflicts. In the current practice, the traffic situation is mostly predicted in a relatively simple manner, such as the parallel-shift prediction method, in which depending on the current position and delay of a train, the subsequent time-distance-line of the train can be parallel-shifted to a later position. This may, however, result in unrealistic results. Traffic prediction on different levels of accuracy has been studied in a few works of research: in [D'Ariano, 2008], fixed running and dwell times are used in the prediction module without consideration of actual traffic situations, while in [Kecman, 2014], the running time and dwell times are dynamically estimated by using the predetermined functional dependence of process time on actual delays.

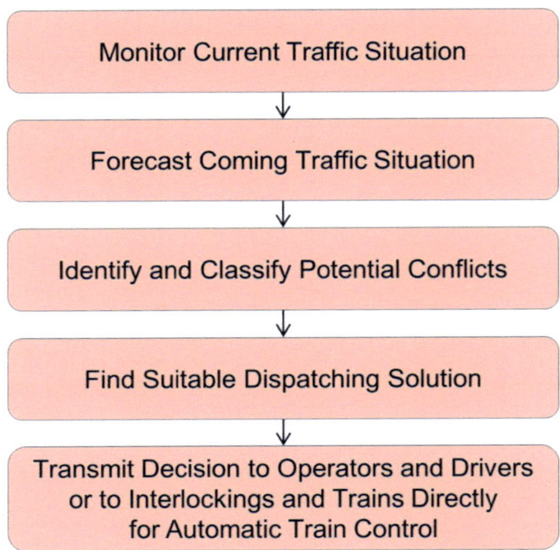


**Figure 2-4: General Dispatching Process**

Once conflicts are identified, the conflict should be classified primarily. Six types of conflicts are defined in [Martin, 1995] as follows:

- Conflicts at track sections or routes
- Conflicts at scheduled stops
- Connection conflicts
- Timetable conflicts
- Dispatching conflicts
- Deadlock conflicts

Conflicts at track sections or routes and scheduled stops belong to occupancy conflicts. This kind of conflict occurs when the requested infrastructure resources of a train are being occupied by another train, or if several trains request the same infrastructure resources concurrently. The resolution of occupancy conflicts will be discussed in detail in Chapter 6. Connection conflicts refer to the risk of transferring not only passengers or goods but also staff, locomotives, cars and coaches from a delayed feeder train to its connected train. Timetable conflicts are caused by the deviation of train movement from the timetable, which may potentially result in conflicts with the other trains. Dispatching conflicts are a result of the inconsistency of the dispatching solutions from different levels or on the same level of the dispatching system. Deadlock conflicts refer to the situation in which several trains are blocked by each other and none of them can move further. Deadlock avoidance has been studied in depth in [Pachl, 1993] [Pachl, 2011] [Cui, 2010]. The general handling procedures for the six types of conflicts were discussed in [Martin, 1995].

The basic methods of railway dispatching can be summarized into two categories: time-related dispatching and location-related dispatching [Martin, 1995]. Time-related dispatching is most commonly used to ensure the punctuality of trains. In the design of an operating schedule, recovery times should be added to the pure running time between scheduled stops and the dwell time at scheduled stops, to enable trains to compensate for minor delays during the operation process. There are two kinds of recovery time for running time [Pachl, 2002]:

- Regular recovery time, which is 3-7% of pure running time and used to compensate the influence of regular train delays.
- Special recovery time, which is a fixed supplement to the pure running time of a concerned section, and is used to compensate for the influence of maintenance or construction works on the concerned section.

Once a train is delayed during the operation process, the train can shorten the running time or dwell time according to the location of the train at that time and predefined recovery times. In addition, time-related dispatching includes not only shortening but also extension of running time and dwell time [Cui, 2010]. For instance, unscheduled waiting time may be designated to the train to be overtaken when performing an overtaking task in a station.

By location-related dispatching, a new train path consisting of a series of reference points will be designated to the dispatched train [Martin, 1995]. A reference point could be a station route, a track or a line section. Location-related dispatching includes overtaking, passing, replatforming, detours, shortening scheduled train paths and so on, and it may be necessary when an infrastructure element (e.g. signal, point or track) is disturbed (e.g. system failure) or unavailable (e.g. unscheduled maintenance) or if the requested infrastructure resources are being occupied by other trains.

Time-related and location-related dispatching can be used independently or be integrated together to solve conflicts. For the resolution of a concrete conflict, the selection of a specific dispatching method or a combination of dispatching methods should depend on the location of the conflict and the surrounding traffic conditions. This will be elaborated in Chapter 6.

## 2.3 Railway Operation Modelling

As the basis of a dispatching assistant tool, special attention should be paid to the railway operation modelling in the design phase. Depending on the level of details of the description, the models can be classified into microscale, mesoscale and macroscale models. The models also can be classified based on the implemented modelling methods. For the sake of clarity, the description level of the model is chosen as the main line of this section, and the dispatching optimization approach (i.e. combination of modelling method and optimization algorithm) will be explained in Section 2.4.

Microscale simulation models have been widely used in practice, including examples such as the software RailSys (by Leibniz Universität Hannover and Rail Management Consultants) [RMCon, 2016], OpenTrack (by Eidgenössische Technische Hochschule Zürich) [OpenTrack, 2016] and LUKS (by VIA Consulting and Development GmbH) [VIA-Con, 2016]. To simulate railway operation processes, two kinds of simulation models are available: synchronous (e.g. RailSys and OpenTrack) and asynchronous (e.g. LUKS). Synchronous models update all train movements simultaneously step by step, and asynchronous models update train movements successively in the timetable according to the priority of the trains [Siefer, 2008]. Restricted by the different processing techniques employed in these two kinds of models, dispatching strategies implemented in asynchronous models are more stringent compared to those implemented in synchronous models (for more comparison of synchronous and asynchronous simulation models it is referred to Section 2.4.2.1). Hence, it has been decided that synchronous models are preferable for the sake of this dissertation.

For the design of simulation models, performance issues should be considered. Within a small study area, the railway operation processes can be simulated precisely with a microscopic model. Within a large study area, the computational complexity on the microscopic level is not yet acceptable, so a simulation model should be designed especially to reduce complexity. There are two main patterns to reduce the computational complexity: decomposition and abstraction.

With decomposition patterns, a network is subdivided into several local areas. Local solutions are generated separately, and then these solutions are coordinated on a higher lever to obtain a globally feasible solution. The key point of decomposition is the coordination between the local solutions. In [Corman et al., 2010; Corman et al., 2011a], a coordination theorem based on an alternative graph model is proposed as constraints on borders of local areas, and the algorithm is tested with examples from different numbers of local areas and with different levels of deviation. The test results show that a globally feasible solution can be difficult to obtain in case of large traffic prediction horizons and severe deviations occur when there are more than nine local areas. Therefore, the best way to coordinate a large number of local areas is still an open question. Due to this reason, an abstraction pattern will be used in this dissertation.

In the case of abstraction patterns, microscopic data are abstracted to the macroscopic level, and macroscale models are capable of covering a large study area. However, macroscale models abstract the whole study area in the same manner, and the simplification of significant subareas may lead to infeasible solutions on the microscopic level. Therefore a multi-scale model is proposed in this dissertation, which includes microscopic, mesoscopic, and macroscopic levels. The concept of a multi-scale model contains two aspects: different levels of details of both the infrastructure and the operating program. The graph theory is a powerful method to model the infrastructure; it is based on an arbitrary location in a railway network defined as a node, and the connection between the two nodes are defined as a link [Radtke, 2014]. In [Cui and Martin. 2011], on the macroscopic level, a node represents a station or junction, and a link represents a line between the stations or junctions. On the mesoscopic level, a node represents a route node, and a link represents a track or track group. On the microscopic level, a node represents a point, crossing or signal, and a link represents a track or block section. Depending on the abstraction level of the infrastructure, the details of the operating program also vary. In [Kettner et al., 2003], the dispatching algorithm and occupation dependencies are simulated precisely on the microscopic level, because running times and minimum headways can be calculated accurately; on the macroscopic level, running times are given for each train group on a certain link, and the minimum headways are determined for each pair of train groups that run on the same link, both of which cannot be adjusted in real-time. In [Kecman et al., 2012], on the macroscopic level, the capacity of a station is assumed to be infinite, and the constraint on open tracks can be abstracted into three levels: trains on the same open track are separated with minimum headways, and their conflicts on this open track and conflicts of trains from different open tracks on merging or intersecting points are ignored; trains on the same open track are separated with a minimum headway, their conflicts on this open track are considered; and conflicts of trains from different open tracks on merging or intersecting points are ignored; all three of these aspects are considered. The constraints on the lowest abstraction level (the last case) are similar to that of the macroscopic model in [Cui, 2010]. In [Kecman et al., 2012], the comprehensive evaluation shows the model with the constraints on the lowest abstraction level can capture train-reordering actions quite well, but the average knock-on delay calculated with this model varies greatly from the values calculated with the microscopic model. Macroscopic models may fail to detect certain conflicts depending on the abstraction level.

In this dissertation, a multi-scale model characterized by continuous scaling will be developed. This type of model was first proposed in [Cui and Martin, 2011]. In this model, core regions whose accuracy is important with respect to rescheduling processes will be described on a microscopic level, and the surrounding areas will be described on mesoscopic and macroscopic levels. The model can migrate continuously between the different abstraction levels depending on the changing significant values of the nodes or links. An important indicator of a significant value is the propagation scope of conflicts, which can be assessed with delay propagation models. Delay propagation in different scopes has been studied in a few research projects: delay propagation in stations was investigated in [Yuan, 2006], and delay propagation in large scopes was researched in [Goverde, 2010] and [Siefer and Radtke, 2006]. Due to the time limitations of this research, and the complexity of the delay propagation algorithm, simplified methods will be employed to assess propagation scopes instead of accurate delay propagation models for practical use. With the multi-scale model solution, accuracy and computational complexity will be well balanced.

## 2.4    Dispatching Optimization of Railway Operation

For the design of a dispatching optimization module, the dispatching objective should be clearly defined initially. Regardless of whether dispatchers or dispatching assistant tools are being considered, a clearly defined dispatching objective is helpful to improve the quality of dispatching solutions. Several dispatching objectives used in recent research will be reviewed in Section 2.4.1. In order to optimize the operation process based on the defined dispatching objective, a suitable dispatching optimization approach should be accordingly designed. A survey of the recent approaches on railway dispatching optimization will be presented in Section 2.4.2.

### 2.4.1    Dispatching Objective

In order to guide dispatchers, four dispatching objectives are defined in [DB NETZ AG, 420.0105] as follows:

- "quickest possible restoration of the regularity or control state in the operation process";
  („schnellstmögliche Wiederherstellung der Planmäßigkeit bzw. des Regelzustandes in der Betriebsdurchführung")
- "ensuring the fluency of operation";
  („Gewährleistung der Flüssigkeit des Betriebes")
- "improvement of the overall punctuality of all trains";
  („Verbesserung der Gesamtpünktlichkeit aller Züge")
- "maximum utilization of the capacity of tracks and nodes".
  („maximale Auslastung der Kapazität von Strecken und Knoten")

The objectives in this guideline are general verbal expressions without quantitative indicators. Different dispatchers may have different understandings of the objectives in a certain situation, and it is difficult to ensure the fulfillment of the objectives in real-time operation. Some researchers have attempted to define objective functions quantitatively, such as the adjusted total malus (the penalty value of train delays) in [IEV, 2011; Cui et al., 2012], the punctuality and fluency of operation in [Martin, 1995; Cui, 2010], and the total knock-on delay in [D'Ariano, 2008]. Additionally, under different dispatching conditions, the importance of objectives may differ. In [Martin, 1995] viscosity[1] is introduced to evaluate dispatching conditions. Punctuality is the primary objective when viscosity is low, and fluency of operation is the primary objective when viscosity is high. In [Luethi et al., 2007], the maximization of productivity is the primary objective when small delays occur, and the maintenance of the circulation plan is the primary objective when a situation causing reduced availability of vehicle or infrastructure occurs. In [Larsen et al., 2013] minimization of the maximum knock-on delay is chosen as the dispatching objective function. In this dissertation, a simplified form[2] of the dispatching objective defined in [Martin, 1995] will be used.

## 2.4.2 Dispatching Optimization Approach

In general the dispatching optimization approaches can be classified into three types: simulative, analytical and heuristic approach according to [Martin, 2002] and [Cui, 2010] (similar classification also can be found in [Corman and Meng, 2013]).

### 2.4.2.1 Simulative Approach

The simulative approach intends to simulate the operation process as in reality. Along with the progress of the simulation process, future traffic situations are predicted iteratively. Once a conflict is detected, the integrated dispatching system will be trigged to solve the conflict. The conflict resolution mechanism depends on the processing technique of the simulation model (synchronous or asynchronous).

In the synchronous simulation model, all train movements are processed simultaneously and interact with each other immediately. The simulation process cannot be rolled back, and the system must have the ability to respond to all kinds of situations immediately [Siefer, 2008]. Therefore, the task of integrated dispatching systems is to imitate the dispatcher to make decisions in case of conflicts. This can be built based on rule-based dispatching. In the simplest manner, the First Come First Served

---

[1] Viscosity: the total knock-on delay of all the conflicted trains divided by the number of infrastructure elements of the dispatching concerned network.

[2] The timetables to be optimized in this dissertation do not include recovery times. For the ease of calculation, one part of the dispatching objective function related to recovery times is excluded (see Section 6.2). The calculation of objective function value is implemented as a separate module, so the objective function can be easily updated or replaced in further applications. This module does not affect the design of the overall optimization algorithm.

(FCFS) dispatching principle can be implemented. In practice, some general dispatching rules have been established in [DB NETZ AG, 420.02], such as "if trains are equivalent, the faster trains always have priority over slower trains". These rules have been more or less implemented in the existing simulation tools. For instance, in the software RailSys several dispatching measures are defined, which include overtaking, replatforming, dwell time extension and so on. A dispatching measure is triggered only if the anticipated delay caused by a conflict exceeds a pre-defined minimum lateness [RMCon, 2007]. The minimum lateness can be manually configured by users, and case specific configuration is necessary to improve the performance of the dispatching module [Marin et al., 2015]. At the end of the simulation, a conflict-free dispatched timetable will be derived.

In the asynchronous simulation model, the trains are inserted in the time-distance diagram successively according to their priorities [Siefer, 2008]. A dispatching assistant tool based on asynchronous simulation called ASDIS (Asynchronous Dispatching) was developed in the research project DisKon [Shaer et al., 2005]. In the ASDIS method, the trains are ranked based on their priorities. The highest-ranking group is inserted initially, and the trains of the same group (with the same priority) ought to be inserted simultaneously. After a group of trains is introduced, conflicts among the equal or higher-ranking trains will be identified and chronologically solved. In the process of conflict resolution, knock-on conflicts are likely to occur, which also need to be added in the conflict list and resolved with the others. Partial priority is applied, which allows a lower-ranked train to take precedence over a higher-ranked train if the delay of the lower-ranked train exceeds a predefined threshold value [Jacobs, 2008]. When all conflicts are resolved, the next group of trains will be inserted. By repeating this process, a conflict-free dispatched timetable will be generated as the final result.

Duo to the different processing techniques, the synchronous and asynchronous simulation approaches have their advantages and disadvantages. Because all train movements are simultaneously processed using the time-step method in synchronous simulation, a deadlock problem may arise during the simulation process, especially on infrastructures with bidirectional operations. Deadlock avoidance algorithms should be additionally implemented to regulate train movement (see [Pachl, 2011] and [Cui, 2010]). This will increase the required computation time. However, owing to the simultaneity of the update of train movements, it is possible to implement more flexible dispatching strategies in the model [Siefer, 2008]. On the contrary, for asynchronous simulation models, the deadlock problem does not exist in general, but the implemented dispatching strategies are more stringent compared to its counterpart. To combine their advantages, the convergence of the synchronous and asynchronous simulation models has emerged in the current research [Jacobs, 2008].

## 2.4.2.2 Analytical Approach

With analytical approaches, train operations are mathematically formulated, and the objective function defines the dispatching strategy (see Section 2.4.1). Recently many mathematical models based on different modelling methods have been developed, such as linear programming (e.g. [Martin, 1995] and [Cui, 2010]), queuing theory (e.g. [Marinov and Viegas, 2011]), and alternative graph model (e.g. [Corman et al., 2011b] and [D'Ariano and Pranzo, 2008]) and so on. For the details of each modelling method it is referred to the aforementioned literatures, and for an overview of mathematical railway dispatching models it is referred to the literature reviews presented in [Alwadood et al., 2012] [Cacchiani et al., 2013] and [Corman and Meng, 2013]. For analytical models, the most challenging task is to find the optimal solution. Due to the strict safety requirements of railway operation, quantities of constraint equations are necessary in an analytical model. Solving such kind of problem with exact methods (e.g. linear programming technique, column generation) is highly complex (even NP-hard (non-deterministic polynomial-time hard)) and time-consuming. Therefore, quite a few of the recent studies have chosen a heuristic approach to solve this problem.

## 2.4.2.3 Heuristic Approach

The heuristic approach is employed to find an approximate solution when solving an optimization problem with exact methods is impractical or impossible. Through the heuristic approach, the solution quality and computational complexity can be well balanced. Examples of the heuristic approach include knowledge-based expert systems, tabu search, simulated annealing, genetic algorithms, swarm intelligence and so on.

The knowledge-based expert system typically consists of two parts – knowledge base and rules engine. The domain knowledge is stored in the knowledge base, and formulated and organized within the rules engine, such as using the IF-THEN logic. When a specific situation occurs, the relevant knowledge is directly withdrawn from the knowledge base according to the pre-defined rules. A fuzzy knowledge-based railway dispatching support system is developed in [Fay, 2000]. The knowledge was acquired at the dispatching control center, and the Fuzzy Petri Net approach is used to establish the rule base. The algorithm is tested on an exemplary case of connection conflicts, and provides promising results. In the project RUDY (Regional Enterprise-spreading Dynamic Sampling of Timetable Information, Reservation and Operation in Public Transport) [Tritschler et al., 2005], an accident-management system for regional public transport is developed based on an expert system. When an accident or traffic jam occurred, the system is capable of suggesting possible detour routes based on the matched pre-recorded historical decisions. The quality of an expert system highly relies on the quality and completeness of the acquired knowledge.

The genetic algorithm is a typical population-based metaheuristic algorithm. In a genetic algorithm, a population of individuals is evolved iteratively through selection, recombination and mutation in order to find a satisfactory solution. The population in each iteration is called a generation, and each individual refers to a candidate solution. In each generation, all candidate solutions are evaluated with a pre-defined objective function, and the best among them will be kept to parent new candidates by a crossover/mutation operator for the next generation. This process is executed iteratively until a termination specification is fulfilled (e.g. a fixed number of generations). The railway dispatching algorithm in [Fan, 2012] is developed based on genetic algorithm. The solution is represented by the sequences of trains on infrastructure resources. The initial population is randomly generated based on the sequences of trains derived from the FCFS principle. Total delay cost is chosen as the dispatching objective function. In each generation, pairs of the best solutions are selected to parent new solutions by a two-point crossover operator. For instance, on a certain track section the sequence of trains is denoted by L1 for Solution 1 and L2 for Solution 2. Two points X and Y are randomly selected in these two lists, and the parts between X and Y in L1 and L2 will be swapped. Accordingly, two new offspring are generated. In the simple testing scenarios, optimal dispatching solutions are found using Fan's algorithm.

Tabu search is a local search-based metaheuristic algorithm. Following the local search procedure, the immediate neighbors of the initial solution are enumerated and evaluated. A neighbor is then selected to replace the initial solution. This process is executed iteratively until certain terminate specification is fulfilled. In order to avoid being trapped in suboptimal regions, an adaptive memory structure is introduced in tabu search. After a solution is visited, it is marked as "tabu" and will be prohibited to be revisited in the next period of time. A macroscopic railway dispatching optimization algorithm based on tabu search is developed in [Cui, 2010]. Total weighted trip time is the dispatching objective function, and changing of train sequences on open track sections are used as the basic move operations for tabu search. Both short term memory (intensification strategy) and long term memory (diversification strategy) are implemented. In [D' Ariano, 2008] another approach is proposed. Tabu search is employed to optimize the paths of trains. After the paths of all trains are selected, the sequences of trains on infrastructure resources will be optimized by the Branch and Bound algorithm. These two processes are executed iteratively in sequence. Based on the previous IEV research, tabu search is preferred as the basis of the dispatching optimization algorithm developed in this dissertation (see Chapter 6).

# 3  Multi-scale Simulation Model

To balance the computational complexity and accuracy of the simulation model, a multi-scale simulation model characterized by continuous scaling is developed in this chapter, in which simulation is concurrently carried out on microscopic, mesoscopic and macroscopic levels. In this model, the significant areas are simulated on a microscopic level, and the other areas are simulated on more efficient mesoscopic and macroscopic level. The processing technique – synchronous simulation – is adopted in this dissertation. In [Cui, 2010] the framework of a synchronous model is developed, and, more significantly, Banker's algorithm is employed to avoid deadlocks in the simulation process. In order to enable the multi-scale simulation model can transform smoothly among different descripted levels, the developed microscopic, mesoscopic and macroscopic model followed CUI's framework of the synchronous model, and the integrated Banker's algorithm was also adopted. The microscopic simulation model will be elaborated in Section 3.1, and the mesoscopic and macroscopic simulation models will be elaborated in Section 3.2. The multi-scale simulation model to be described in this chapter was developed within the framework of the DFG project [Martin and Liang, 2017].

## 3.1  Microscopic Simulation Model

In the modelling of railway systems, both the structural and the behavioral perspectives should be considered. The structural perspective is concerned with the fundamental components constructing the simulation model (Section 3.1.1), and the behavioral perspective with the workflow of the simulation model, which reflects the interaction mechanism between the components (Section 3.1.2).

### 3.1.1  The Components of Synchronous Simulation

The fundamental components of the synchronous simulation model include infrastructure resources, simulation performers and simulation tasks, which, in reality, correspond to infrastructure, trains and timetables of railway operation.

#### 3.1.1.1  Infrastructure Resources

To depict a railway network mathematically, the link and node model established in [Radtke, 2014] was used, which has been proven to be a powerful tool to describe complex railway infrastructure networks in practice. A node is defined as an arbitrary location in a railway network, and a link as a connection between two nodes. Nodes are differentiated into various types (signals, points, timing points etc.), and links store all relevant information (e.g. speed, gradient, radius etc.). The link and node model will not be elaborated herein, for further details it is referred to [Radtke, 2014; RMCon, 2007].

To prepare the input data for the simulation model, a given node and link mode depicted infrastructure network is decomposed into many individual infrastructure resources according to the method developed in [Martin et al., 2012]. An infrastructure resource is defined as a basic structure, which is the maximum occupation unit allowed to be occupied by only one train simultaneously on a microscopic level. Basic structures are requested, allocated and released as a basic unit in the simulation process. The boundaries of a basic structure could be block signals, signal clearing points or route clearing points[3]. The software PULEIV developed by IEV (Institut für Eisenbahn und Verkehrswesen der Universität Stuttgart) [Martin et al. 2008a; Martin et al. 2008d; Martin et al. 2008c] can be used to decompose an infrastructure network automatically into its basic structures. An example of basic structures partitioning is shown in Figure 3-1. As it is seen in the figure, a turnout is depicted by the node and model with five nodes and the links between them. Because the boundaries of the turnout (i.e. node 1, node 3 and node 5) are either signal clearing points or route clearing points, the turnout is divided into a basic structure directly. After the basic structure is established, the internal nodes (i.e. node 2 and node 4) will be removed, and new edges will be created to replace the existing links. An edge is defined as a connection between two boundaries of a basic structure. An edge can also be regarded as a combination of a series of links in the same direction. The attributes of a new edge including length, permissible speed, gradient and radius can be deduced from the included links. There are two categories of basic structures: junction-type and non-junction-type basic structure. The basic structures, containing turnouts

---

[3] In [Pachl, 2016] a block signal is defined as "a signal that governs train movements into a block section"; a signal clearing point is defined as "the point at the end of the overlap a train must have cleared completely to release the block section in approach of the signal"; a route clearing point is defined as "A point that a train must have cleared completely before a locked route or sections of a locked route may be released".

or crossings, belong to the junction-type (the basic structure marked in green in Figure 3-1), and the others, containing only tracks, belong to the non-junction-type (the basic structure marked in light red in Figure 3-1).

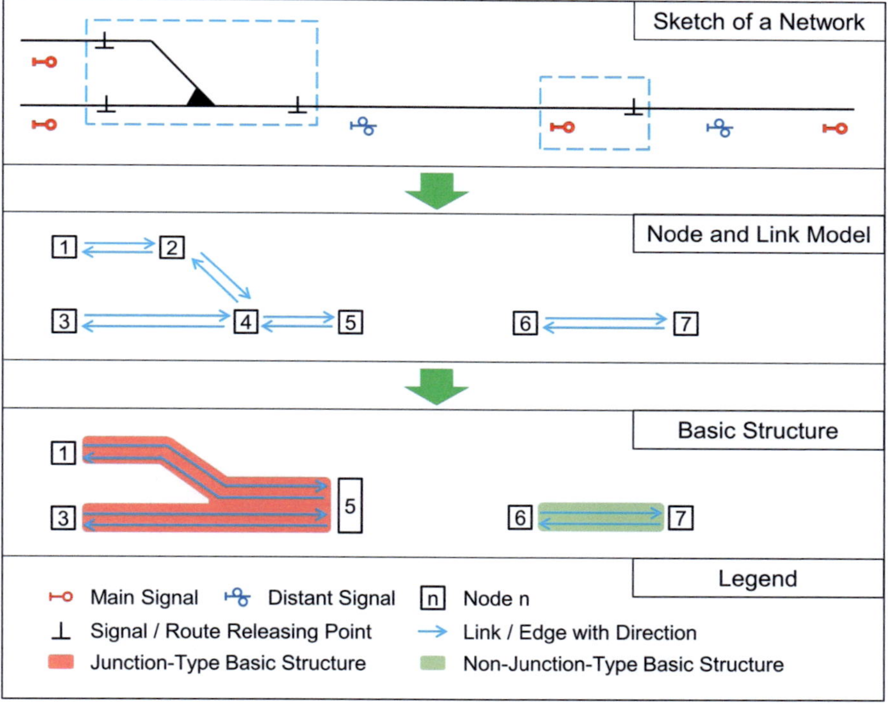


**Figure 3-1: An Example of Basic Structure Partitioning Algorithm (source: [Martin and Liang, 2017])**

The prepared input data on edges and basic structures should include the following information:

**Edge Attributes**

−   Edge ID (an unique integer identifier)

−   Starting node ID and end node ID

−   Length [m] (sum of the lengths of the included links)

- Permissible speed [km/h] (the lowest speed limit among the included links[4])
- Gradient [‰] (positive for uphill and negative for downhill, weighted average radius of the included links)
- Radius [m] (weighted average radius of the included links)

**Basic Structure Attributes**

- Basic structure ID (an unique integer identifier)
- IDs of the included edges (at least two edges, one in each direction)
- Whether, it is a free resource[5] [true or false]

After the edges and basic structures of an infrastructure network have been determined, the input data on block sections can be prepared. A block section is comprised of a series of edges in sequence, and the corresponding basic structures are also included as an attribute of the block section. Moreover, the automatic train protection (ATP) system of the block section should be designated. In practice, information about movement authorities and speed limits is transmitted by ATP systems between track and train, in order to trigger automatic braking if valid limits (e.g. stop signals and speed limits) are violated. In the simulation model, the ATP system is used to control train movement, which will be elaborated in Section 3.1.2. There are two kinds of ATP systems:

- Intermittent ATP,
- Continuous ATP.

In the intermittent ATP data is transmitted at discrete points located along the track. A simple form of the intermittent ATP system is implemented within this approach, which only takes the main signal and distant signal into consideration as the data transmission points. The block section equipped with the intermittent ATP is bounded by two main signals. In the case of continuous ATP the data is transmitted continuously through cab signaling system, and block markers installed along the trackside

---

[4] The permissible speed on an edge is conservatively calculated. Nevertheless, other calculation methods can also be applied. For instance, the weighted average speed limit of the included links is also an appropriate option.

[5] A free resource is defined as a virtual resource extended from the investigated infrastructure network for the convenience of modelling, which any train is free to enter at any time (no need of movement authority).

are used to indicate the boundaries of the block sections. In the transition zone from the intermittent ATP (continuous ATP) to the continuous ATP (intermittent ATP), the boundaries of a block section could be a main signal and a block marker, with such a block section also belonging to the block section equipped with continuous ATP. As an example, block sections equipped with the two different ATP systems are shown in Figure 3-2.

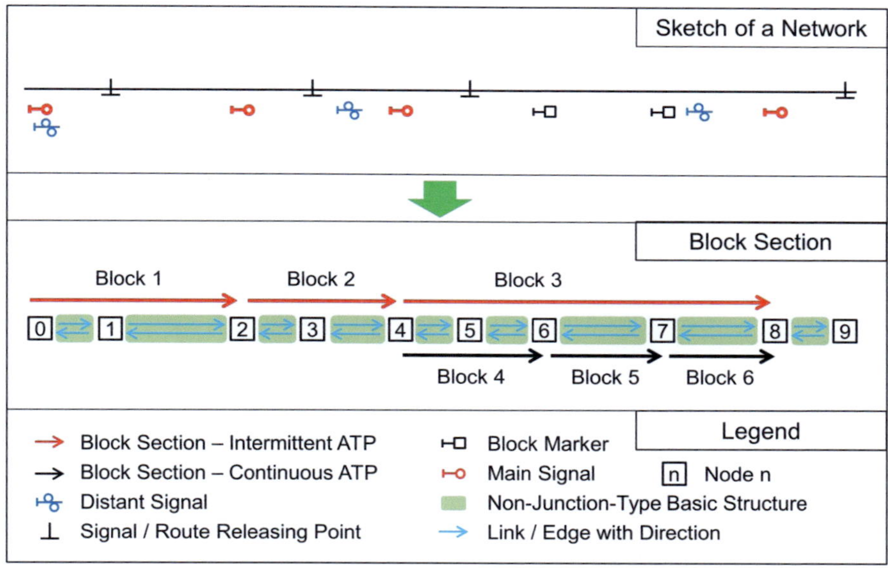


**Figure 3-2: Example of Block Sections (modified from [Martin and Liang, 2017])**

In the intermittent ATP system, the main signal only provides information about the block section behind the signal, and the distant signal gives the approach information for the next signal, as shown in Figure 3-3. The position of the distant signal influences the control of train movements (see Section 3.1.2). In the simulation model, the distant signal is not expressed as a concrete infrastructure node, but rather as the distance from itself to the entrance signal of the block section in which it is physically located (shown in bold type in Figure 3-3). On railway lines with short block sections where the distant signal and the entrance block signal are located at the same location (e.g. main signal 9 and distant signal D11 in Figure 3-3), the distant signal distance should be set to 0.

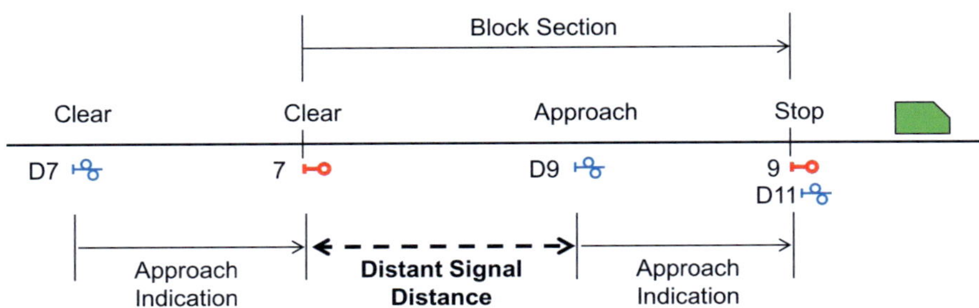


**Figure 3-3: Calculation of Distant Signal Distance (modified from [Martin and Liang, 2017])**

For the block section belonging to intermittent ATP, the block overlap is also modelled as an attribute of the block section. An overlap consists of a series of basic structures (e.g. Overlap 1 and Overlap 2 in Figure 3-4), and is identified by the block section which the overlap belongs to operationally (e.g. Block 1 in Figure 3-4) and the block section where the overlap is physically located[6] (e.g. Block 2 or Block 3 in Figure 3-4). Thus, depending on the path of a train (e.g. Train 1 runs towards Block 2 or Block 3 in Figure 3-4 ), the corresponding overlap could be chosen. Furthermore, on railway lines with block overlaps, the control length of a main signal includes the block section and the corresponding overlap (e.g. control length of signal 7 in Figure 3-4), which means that the block section and the corresponding overlap should be requested as a whole in the simulation process, and that the train movement into the block section is only authorized when all basic structures in the control length are released.

---

[6] In this approach, one overlap is considered for one running direction. In special cases that a block section has several overlaps in various lengths for different situation dependent speeds in a running direction, an additional identification attribute about train types should be introduced for the overlap, besides the two attributes about block sections as mentioned above.

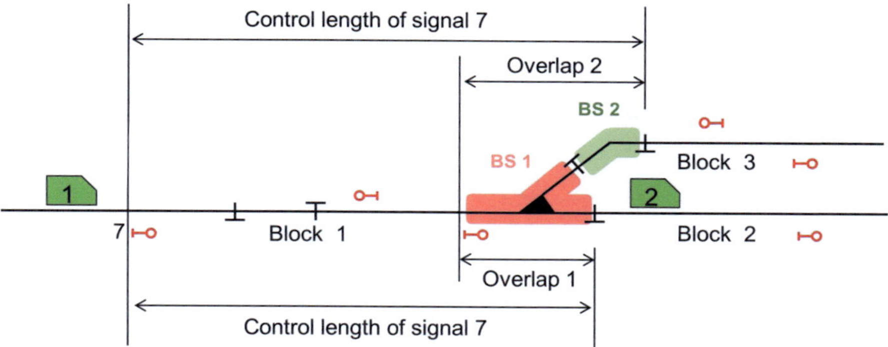



**Figure 3-4: Example of Overlaps**

The prepared data on block sections should include the following information:

## Block Section Attributes

- Block section ID (an unique integer identifier)
- IDs of the included edges and corresponding basic structures
- ATP system (intermittent or continuous)
- Distant signal distance (in case of intermittent ATP system)

In case the block sections of an infrastructure network are given, train paths can be determined. A train path consists of a sequence of block sections that guide a train run through the infrastructure network. In particular, the first and last block sections in the sequence always contain exclusively free resources, where train movements start and terminate. Furthermore, although the trajectories of two train paths may be identical, the train paths may not be treated as the same if they are composed of different block sections. As an example two train paths are determined based on a given network sketch in Figure 3-5, and train path 1 can be used by trains compatible with intermittent ATP system and train path 2 can be used by train compatible with continuous ATP system. The prepared data on train paths should include the following information:

## Train Path Attributes

- Train path ID (an unique integer identifier)
- IDs of the included block sections

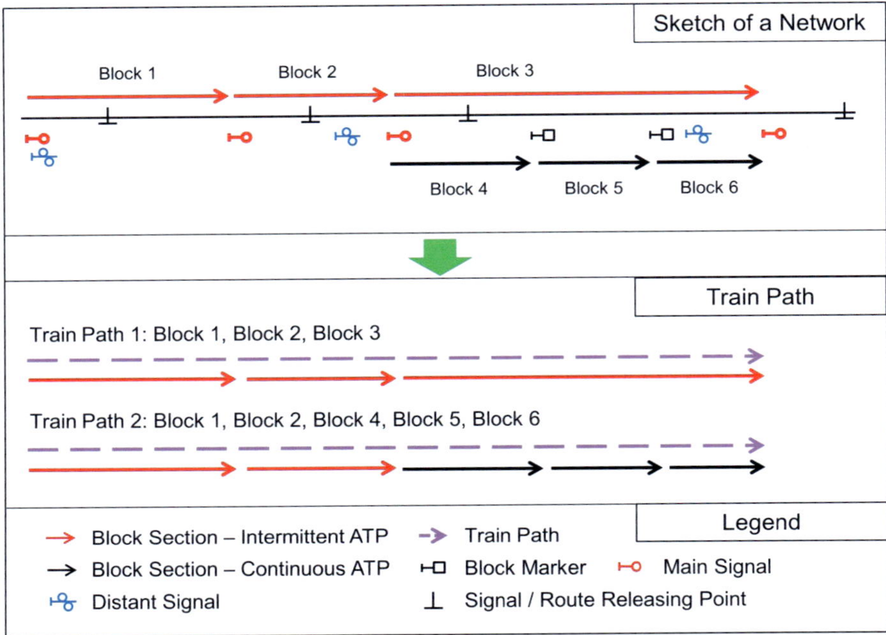


**Figure 3-5: Example of Train Paths (modified from [Martin and Liang, 2017])**

With edges, basic structures, block sections and train paths any given infrastructure network can be described in the simulation model, and the data will be loaded at the beginning of a simulation.

## 3.1.1.2 Simulation Performers

A simulation performer refers to a train in the simulation model. A train is defined as a locomotive or self-propelled vehicle, along with or coupled to one or more vehicles with the authority to operate on main tracks in accordance to rules specified for train movements [Pachl, 2002]. In this dissertation, only train movements are considered, and shunting movements are not taken into account. In the simulation workflow described in Section 3.1.2, a train serves as a basic entity to request, occupy and release infrastructure resources (basic structures), and the requesters and occupiers of an infrastructure resource will be updated at a certain time interval according to the simulation logic. At the beginning of each simulation, the characteristics of the investigated trains should be loaded, which include the following attributes:

- Train ID (an unique integer identifier)
- Train production type (passenger train or freight train)
- Maximum speed of the train [km/h]
- Mass of the traction unit [kg]
- Length of the traction unit [m]
- Mass of a vehicle [kg]
- Length of a vehicle [m]
- Number of vehicles
- Braking acceleration [m/s$^2$]
- Compatible ATP system (intermittent or continuous)
- Parameters of the traction unit resistance
- Tractive effort - speed diagram [N – km/h]

Based on these attributes of trains and the relevant attributes of the infrastructure, the train dynamics, especially the acceleration phase, can be described, which is fundamental for the estimation of the forward distance of a train in one time interval (see Appendix I). The fundamental equation of running dynamics is shown in Formula (3-1) (see also [Wende, 2003]).

$$F_{Tr}(v) - F_R(v) = \rho \cdot m \cdot a_{Tr} \tag{3-1}$$

Notation used:

$F_{Tr}(v)$:     Tractive effort at wheel at a given velocity $v$ [N]

$F_R(v)$:     Train resistance at a given velocity $v$ [N]

$\rho$:     Coefficient of increase in mass [-]

$m$:     Mass of the train [kg]

$a_{Tr}$:     Acceleration rate [m/s$^2$]

The tractive effort at the wheel generates the power to accelerate the train, and is normally depicted by a tractive effort – speed diagram as shown in Figure 3-6. The tractive effort – speed diagram can be described with a set of hyperbolic or parabolic formulas and each of them is defined only for a certain speed interval (the detailed description can be found also in [ Brünger and Dahlhaus, 2014; Quaglietta, 2011]).

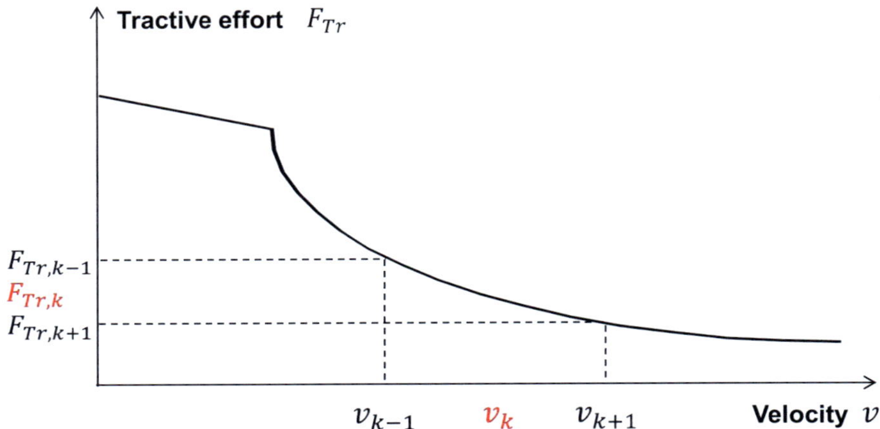


**Figure 3-6: Tractive Effort – Speed Diagram**

Moreover, the tractive effort – speed diagram can also be described with a series of support points, where each support point refers to a concrete speed and the corresponding tractive effort [RMCon, 2007]. The support points are arranged very closely, so the curve between two points can be approximated by a straight line. An example of the support points is shown in Table 3-1.

| From $v$ [km/h] | From $F_{Tr}$ [N] | To $v$ [km/h] | To $F_{Tr}$ [N] |
| --- | --- | --- | --- |
| 0.000 | 98885 | 4.600 | 98875 |
| 4.600 | 98875 | 9.200 | 98855 |
| ... | ... | ... | ... |
| 96.000 | 21704 | 100.00 | 21386 |

**Table 3-1: Example of Support Points of Tractive Effort-Speed Diagram**

The second description method was implemented in this simulation model. As an example, the procedure to determine the tractive effort $F_{Tr}(v_k)$ at a given speed $v_k$ is shown in Figure 3-6. Firstly, the two immediately adjacent support points $v_{k-1}$ and $v_{k+1}$ should be determined, and the corresponding tractive efforts $F_{Tr}(v_{k-1})$ and

$F_{Tr}(v_{k+1})$ can then be read. Next, the tractive effort $F_{Tr}(v_k)$ can be calculated using the linear interpolation method:

$$F_{Tr}(v_k) = F_{Tr}(v_{k-1}) \cdot \frac{v_{k+1} - v_k}{v_{k+1} - v_{k-1}} + F_{Tr}(v_{k+1}) \cdot \frac{v_k - v_{k-1}}{v_{k+1} - v_{k-1}} \qquad (3\text{-}2)$$

The train resistance at a given speed $F_R(v)$ includes traction unit resistance, vehicle resistance and line resistance. For a traction unit (including multiple units), the resistance is normally calculated with the following formula [Brünger and Dahlhaus, 2014]:

$$F_{Rt}(v) = g \cdot m_T \cdot (a_0 + a_1 \cdot v) + a_2 \cdot v^2 + a_{2r} \cdot v_r^2 \qquad (3\text{-}3)$$

Notation used:

$F_{Rt}(v)$:       Traction unit resistance at a given speed v [N]

g:       Earth gravity constant 9.81 m/s$^2$

$m_T$:       Mass of traction unit [kg]

v:       Speed of the train [km/h]

$v_r$:       Relative speed between air and the train [km/h] (i.e. 10 km/h)

$a_0, a_1, a_2, a_{2r}$:       Parameters of traction unit resistance

The Formula (3-3) can also be transformed into a simplified form (3-4) [RMCon, 2007], which was the form implemented in the simulation model.

$$F_{Rt}(v) = a + b \cdot v + c \cdot (v + v_r)^2 \qquad (3\text{-}4)$$

Notation used:

$a, b, c$:       Parameters of traction unit resistance

Regarding vehicle resistance, the calculation methods for passenger trains and freight trains are different. For passenger trains, Sauthoff's formula is used [Sauthoff, 1932]:

$$F_{Rwp} = (1.9 + c_b \cdot v) \cdot \frac{m_w \cdot g}{1000} + 0.047 \cdot (n_w + 2.7) \cdot A_f \cdot (v + v_r)^2 \qquad (3\text{-}5)$$

Notation used:

$F_{Rwp}$:       Vehicle resistance for passenger trains [N]

$c_b$:       Coefficient for the number of axles (0.0025 for vehicles with 4 axles)

$m_w$:          Mass of all vehicles [kg] (sum of the mass of each vehicle)

$n_w$:          Number of vehicles

$A_f$:          Cross-sectional area of the vehicles [m$^2$] (assumed as 1.45)

For freight trains, Strahl's formula [Strahl, 1913] is used to approximate the vehicle resistance:

$$F_{Rwf} = \frac{m_w \cdot g}{1000} \cdot (c_a + (0.007 + c_m) \cdot \frac{(v + v_r)^2}{100}) \tag{3-6}$$

Notation used:

$F_{Rwf}$:          Vehicle resistance for freight trains [N]

$c_a$:          Coefficient for axle adhesion (1.4 for roller bearings)

$c_m$:          Value for air resistance (0.04 for closed wagons)

Line resistance is mainly caused by the gradient of lines[7]. The grade resistance of a train is approximately calculated by:

$$F_{Rlg} = m \cdot g \cdot n \cdot 1000 \tag{3-7}$$

Notation used:

$F_{Rlg}$:          Grade resistance of a train [N]

$m$:          Mass of the whole train

$n$:          Gradient [‰]

Eventually, the tractive effort and the resistances can be integrated into the Formula (3-1) to calculate the acceleration rate of a specific train:

$$F_{Tr}(v_k) - F_{Rt}(v) - F_{Rw} - F_{Rlg} = \rho \cdot m \cdot a_{Tr} \tag{3-8}$$

---

[7] The curve resistance is relatively small, so it is left out in this approach. The detailed calculation method can be found in [Pachl, 2002; Brünger and Dahlhaus, 2014].

The effort represented by the left-hand side of the Formula (3-8) cannot be fully utilized by a train, because the rotating parts of the train will consume some of the effort. This phenomenon is described by the coefficient of increase in mass $\rho$. Furthermore, the coefficients for the traction unit and vehicle are different due to the different physical characteristics, from which the coefficient for the whole train can be derived:

$$\rho = \frac{(\rho_T \cdot m_T + \rho_W \cdot m_W)}{m_T + m_W} \tag{3-9}$$

Notation used:

$\rho_T$:  Coefficient of increase in mass for a traction unit

$\rho_W$:  Coefficient of increase in mass for a vehicle

To calculate deceleration rate, it is only necessary to replace the tractive effort with a braking force (Formula (3-10)). The quantity of braking force is negative, because the direction of braking force is opposite to the train running direction.

$$F_{Br} = \rho \cdot m \cdot a_{Br} \tag{3-10}$$

With

$a_{Br}$:  Braking acceleration rate [m/s$^2$]

In this section, the basics of train running dynamic are elaborated, which will be used to estimate the forward distance of a train in one time interval under the operational and infrastructure-related constraints elaborated in Appendix I.

### 3.1.1.3  Simulation Tasks

A simulation task describes the actions to be performed by a train during operation, and is derived from a pre-provided timetable. A given simulation task is always bound to a certain train, and all the movements of the train in the investigated area are defined as one simulation task. A simulation task could include the following information:

−  Available train paths for the included train
−  Scheduled stops along each train path
−  Scheduled departure time and dwell time at each stop

–    Specific scheduled stop for turnaround (only in case the train is scheduled to turn around at the stop)

With regards to the departure times at scheduled stops, because the minimum running time is implemented in this approach, only the departure time at the initial stop in the investigated area or at the boundary of the investigated area is necessary. The departure times at the subsequent stops along the train path can be easily deduced.

A simulation task has three states - created, running and terminated - in its entire lifecycle. At the beginning of a simulation, the simulation tasks will be loaded as the infrastructure resources and simulation performers. The state of each simulation task is initialized as "created" at this moment. During the simulation process, when the execution time exceeds the scheduled departure time of a train at the initial stop or the boundary of the investigated area, the state of the corresponding simulation task will be changed into "running". Only the simulation tasks with the state of "running" are allowed to be performed in the simulation process. When the whole train has left the investigated area (the train is completely located on the free resource at the end of the train path), the state of the corresponding simulation task will be changed into "terminated". The terminated simulation task will not be considered in the further simulation process.

The infrastructure resources, simulation performers and simulation tasks serve as the structural basis of the simulation model, and they interact with each other during the simulation process. The interaction mechanism will be elaborated in Section 3.1.2.

3.1.2    The Workflow of Synchronous Simulation

The workflow of synchronous simulation developed in [Cui, 2010] is adopted in this dissertation, and is shown in Figure 3-7. In the workflow, a synchronous simulation is represented as a series of single processing steps triggered by a certain time interval, and three activities - request resources, allocate resources and proceed with simulation tasks – are to be conducted in each step. A single processing step is executed iteratively at each time interval along with the increasing accumulated execution time. At the beginning of each single processing step, the terminate specification will be checked. If all simulation tasks are accomplished, the simulation process will be terminated. In [Cui, 2010] several terminate specifications are enumerated, for instance, that a simulation process terminates if a predefined time period is exceeded. In this approach only the aforementioned terminate specification is implemented. At the end of a simulation, important information recorded during the simulation process could be outputted as protocols, such as time-distance diagrams and blocking times for infrastructure basic structures, which are basic data for further result evaluation or optimization.

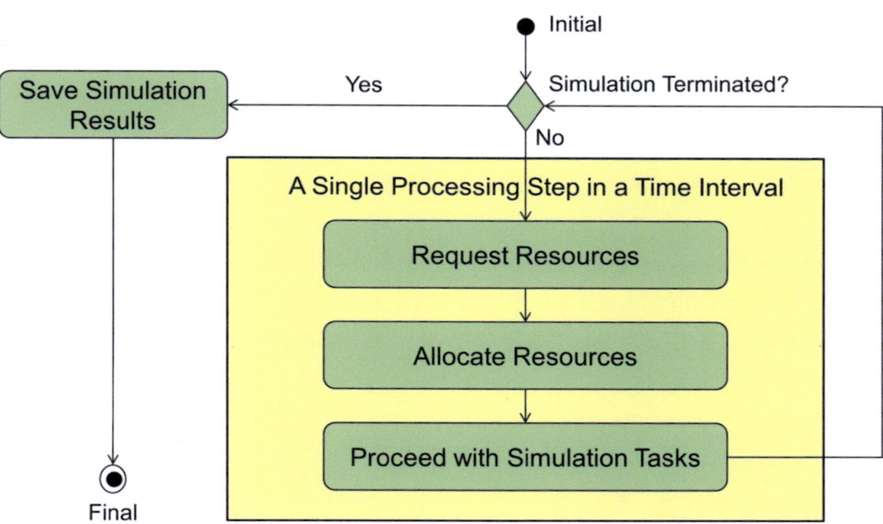


**Figure 3-7: The Workflow of Synchronous Simulation [Cui, 2010]**

## 3.1.2.1    Request Resources

In a single processing step, the resources to be required in the current time interval should be determined. The resource requirement of a simulation task is dependent on the current ATP system, action, physical position and speed of the considered train.

The current ATP system refers to the intermittent or continuous ATP system introduced in Section 3.1.1.1. Regarding the action of a train, four actions including "Run", "Pre-stop", "Stop" and "Turnaround" are defined to support the modelling of train movements with different characteristics. For each simulation task, the action of the included train is initialized to "Run" at the beginning of a simulation. Only if the train has obtained a block section designated as a scheduled stop, can the action be changed into "Pre-stop". At the moment the train is successfully stopped at the scheduled stop, the action should be changed into "Stop", and the dwell time will start to be counted. After the scheduled dwell time is fulfilled, the turnaround task may be executed if necessary, and the action will be changed into "Turnaround"; otherwise the action should be changed into "Run". The "Turnaround" action implies the completion of a turnaround task, and that the corresponding train is ready for departure. It can be seen that only the trains with actions of "Run" and "Turnaround" have the chance to request new resources in one time interval. Moreover, the current position and speed of the train should also be considered to determine the resource requirement in the current time interval.

Eventually, the situations in which the resource requirement is necessary are summarized into 8 cases, and in the other cases, there is no need of resource requirements. Because these 8 cases are defined for resource requirement, a certain Case X will be named as Request-Case X in the following context. As shown in Figure 3-8, only the simulation tasks whose state is "Running" is allowed to be executed in the current time interval, and the 8 request-cases are divided into two groups according to the current ATP system of the involved trains. The two groups of request-cases will be elaborated separately in the following context.

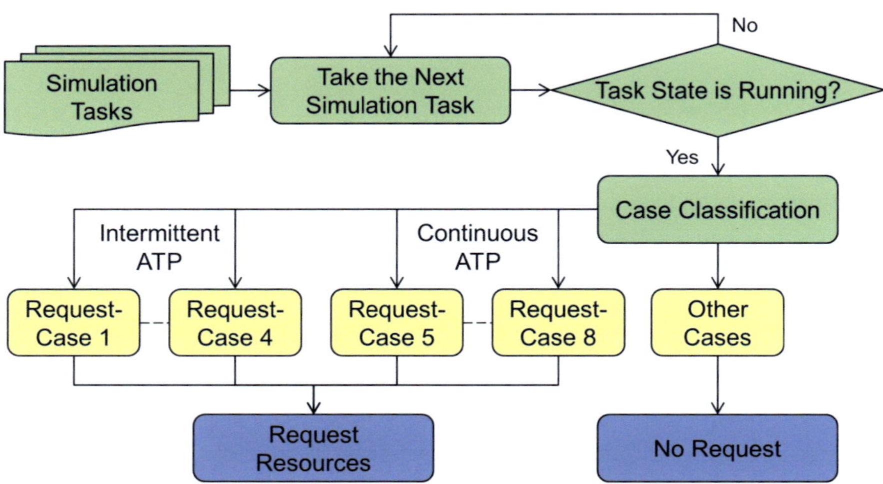


**Figure 3-8: Process of Resource Requirement (source: [Martin and Liang, 2017])**

For the first group of request-cases (Request-Case 1 to Request-Case 4), the involved trains are regulated by intermittent ATP. In Request-Case 1, the train speed is not equal to zero, and action of the train is "Run". The current position of train head is located between the entrance signal of the last current block section[8] and the dis-tance signal for the next block section[9] as shown in Figure 3-9. In the current time interval, the train head will pass the distance signal for the next block section. The forward distance of the train in the current time interval can be estimated by the method that has been described in Appendix I. The required resources include the basic structures covered in the next block section and the overlap, if the next block section still belongs to intermittent ATP system; otherwise only the basic structures covered in the next block section are included. In all 8 request-cases with resource requirements, the simulation model can automatically detect the ATP system used by the new required block section and decide whether overlaps should be included. Therefore, the issue of overlaps will not be particularly explained in the following con-text.

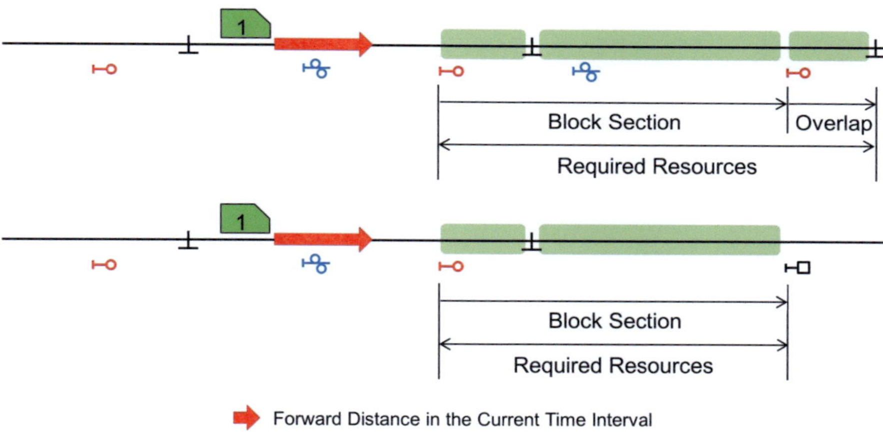



**Figure 3-9: Resource Requirement – Request-Case 1 (source: [Martin and Liang, 2017])**

In Request-Case 2, the train speed is not equal to zero, and the action of the train is "Run". The current position of the train head is located between the distance signal

---

[8] The block sections physically occupied by a train at a certain time instance are defined as the current block sections for the train, where the last section corresponds to train head location and the first to train rear location. For the train completely located in one block section, the first and last current block sections are the same.

[9] The next block section or subsequent block sections in this approach are always counted from the last current block section.

for the next block section and the rear signal of the last current block section, as shown in Figure 3-10. In the current time interval, the train head will pass the distance signal for the after-next block section. The basic structures covered in the after-next block section are requested. In this request-case, the next block section must be absolutely occupied by the train, otherwise the train should decelerate in the current time interval and eventually stop before the rear signal of the last current block section, which means the passage of the distant signal for the after-next block section cannot occur. The resources already occupied by the train will not be requested again in the resource requirement process.

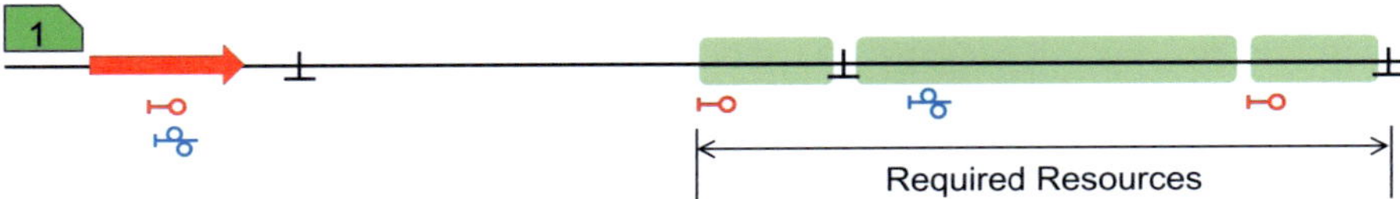


**Figure 3-10: Resource Requirement – Request-Case 2**

In Request-Case 3, the train speed is equal to zero, and the action of the train is "Run". It means that the train is performing an unscheduled stop or has just finished a scheduled stop in front of a main signal as shown in Figure 3-11. The required resources are the basic structures covered in the next block section.

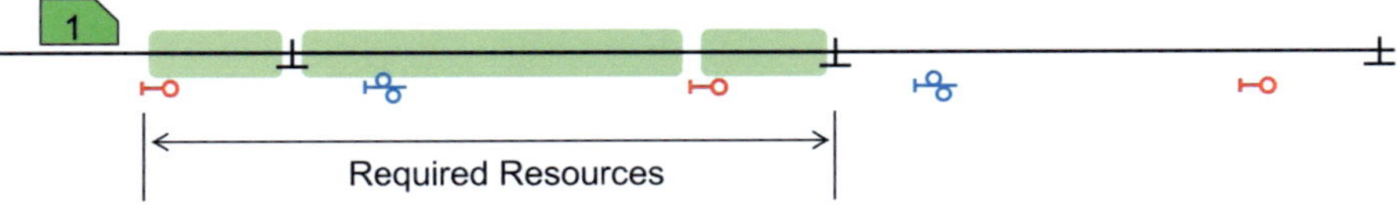


**Figure 3-11: Resource Requirement – Request-Case 3**

On lines with short block sections, the entrance signal of a block section and the distant signal for its next block section can exist at the same location. An example is shown in Figure 3-12: A train (Z1) is performing an unscheduled stop directly in front of the entrance signal of a block section (Block 1). Following the aforementioned resource requirement procedure for Request-Case 3, Z1 should request Block 1 in the current time interval. If the required resources are successfully allocated to Z1, the train head position should be updated depending on the estimated forward distance (marked in red in Figure 3-12). At the end of the current time interval, the train head has fully passed the distant signal for Block 2. As a result, Block 2 will be omitted in the resource requirement process in the next time interval.

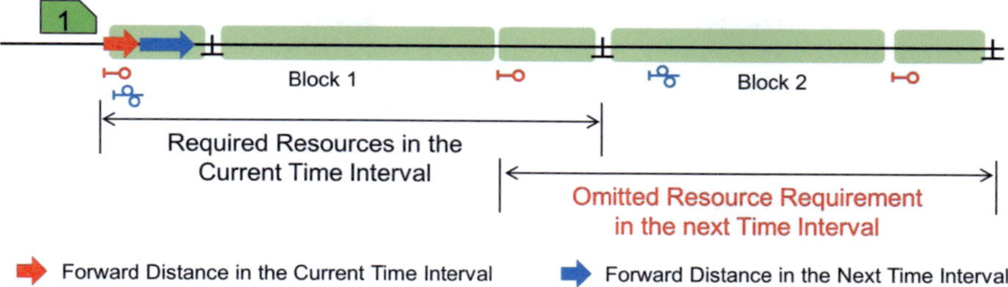



**Figure 3-12: Stopping Point Adjustment (I)**

In order to ensure the logical integrity, the stopping point of the train is adjusted. The new stopping point is not located directly in front of the signal, but maintains a certain distance from the signal as shown in Figure 3-13. The distance is calculated as the forward distance in one time interval when the concerned train reaccelerates after the unscheduled stop (marked in red in Figure 3-13). After the stopping point adjustment, the train head will pass the distant signal in the next time interval (marked in blue in Figure 3-13), and the resource requirement will certainly not be omitted. One time interval is taken as 1 second in this simulation model, and in most cases the maximum acceleration rate of a train is less than 1 m/s$^2$. The position of the stopping point is moved less than 0.5 meters, and the influence of the adjustment is negligible.

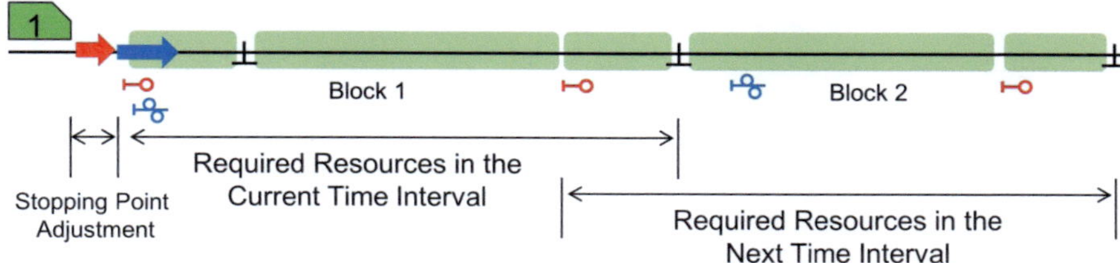



**Figure 3-13: Stopping Point Adjustment (II)**

In Request-Case 4, the action of the train is "Turnaround". A special feature of this request-case is that some basic structures beyond the train head in the last current block section may not yet be occupied by the train, which should also be included in the required resources besides the basic structures belonging to the next block section as shown Figure 3-14.

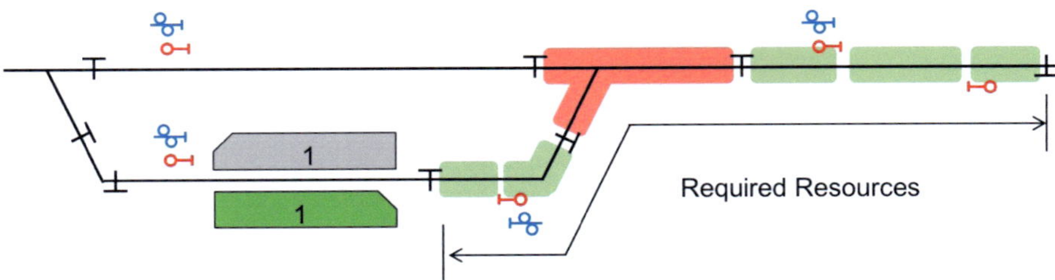


**Figure 3-14: Resource Requirement – Request-Case 4**

For the second group of request-cases (Request-Case 5 to Request-Case 8), the involved trains are regulated by continuous ATP. In Request-Case 5, the train speed is not equal to zero, and the action of the train is "Run". Moreover, the estimated position of the train head at the end of the current time interval will not exceed the rear signal or block marker of the last current block section, or the ATP system of the next block section is not intermittent ATP. In Request-Case 5 only the resource requirement process in continuous ATP territory is considered, and the introduction of the additional constraint is intended to distinguish Request-Case 5 from Request-Case 6, which describes the resource requirement process in the transition zone between continuous ATP and intermittent ATP territory.

In addition to the three constraints above, the nominal stopping point should pass a block marker in the current time interval as shown in Figure 3-15. The required resources are the basic structures belonging to the block section behind the block marker. A nominal stopping point is different from a real stopping point (e.g. a stop signal), and a train does not have to stop before a nominal stopping point. If the specific data of a train (i.e. current position, speed and other train attributes) is provided, the braking curve of the train can be calculated (e.g. the curves marked in red in Figure 3-15). The endpoint of the braking curve is defined as the nominal stopping point.

In order to determine the movement of the nominal stopping point, two braking curves should be calculated. As shown in Figure 3-15, the first braking curve is calculated based on the assumption that the train starts to brake at the beginning of the current time interval, while the second braking curve is calculated based on the assumption that the train starts to brake at the end of the current time interval. If the nominal stopping point is moved from one block section to another block section in the current time interval, new resource requirement is necessary to ensure safe train separation.

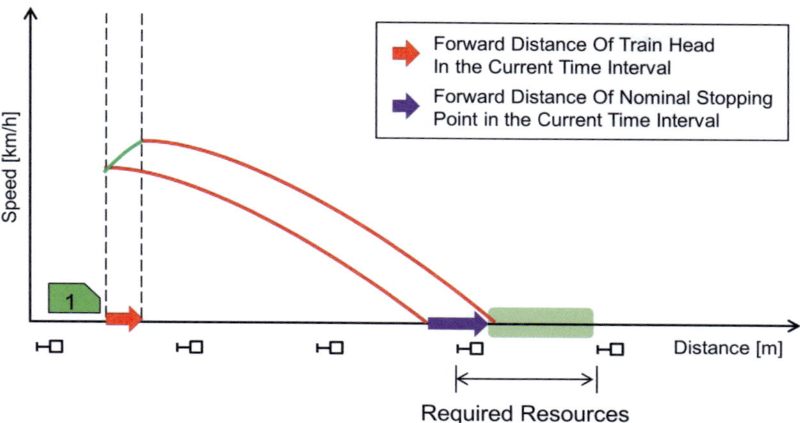


**Figure 3-15: Resource Requirement – Request-Case 5 (source: [Martin and Liang, 2017])**

In Request-Case 6, the current train speed is not equal to zero, and the action of the train is "Run". The train head will pass the rear signal of the last current block section, and the intermittent ATP system is used in the next block section. An example of the transition zone between the two ATP territories is demonstrated in Figure 3-16. If the train head will pass the distant signal for the after-next block section, the basic structures included in the after-next block section should be required.

In Request-Case 7, the current train speed is equal to zero, and the action of the train is "Run", which indicates that the train is performing an unscheduled stop or has just finished a scheduled stop in front of a block signal or block marker. The basic structures included in the next block section should be required in the current time interval. In Request-Case 8, the action of the train is "Turnaround". The required resources should include the basic structures beyond the train head in the last current block section and the basic structures in the next block section. In principle, the processes of resource requirement in Request-Case 7 and Request-Case 3 are the same, and processes in Request-Case 8 and Request-Case 4 are the same.

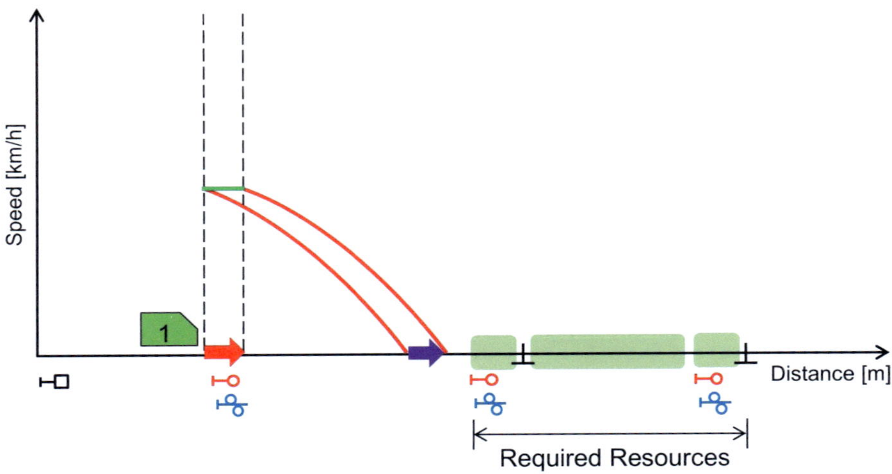


**Figure 3-16: Resource Requirement – Request-Case 6**

## 3.1.2.2    Allocate Resources

Following the procedure described in the previous section, the required resources can be determined for each train. The trains with new resource requirements will be selected, and these resource-requester pairs will be checked in sequence until the last pair is checked. The process of resource allocation is illustrated in Figure 3-17. It can be seen that only if a resource-requester pair successfully passed the conflict-free test and the deadlock-free test, can resource allocation be executed. In the conflict-free test, if one of the required resources[10] is occupied by another train, the resource requirement will be ignored in the following deadlock-free test. Regarding the deadlock-free test, the deadlock avoidance algorithm which is originally developed in [Cui, 2010] is used. The algorithm is designed based on Banker's algorithm, and is intended to avoid deadlocks in railway synchronous simulation. For more details of the algorithm, see Chapter 4 in [Cui, 2010].

After the resource allocation in the current single processing step is completed, all resource-requester pairs should be cleared. The lifecycle of the resource-requester

---

[10] The basic structure belonging to free resource is a special case. According to the definition (see Section 3.1.1.1.), this kind of basic structure is always conflict-free, even though it has been occupied by the other trains.

pairs is only one time interval, because the resource requirement in the next time interval likely varies.

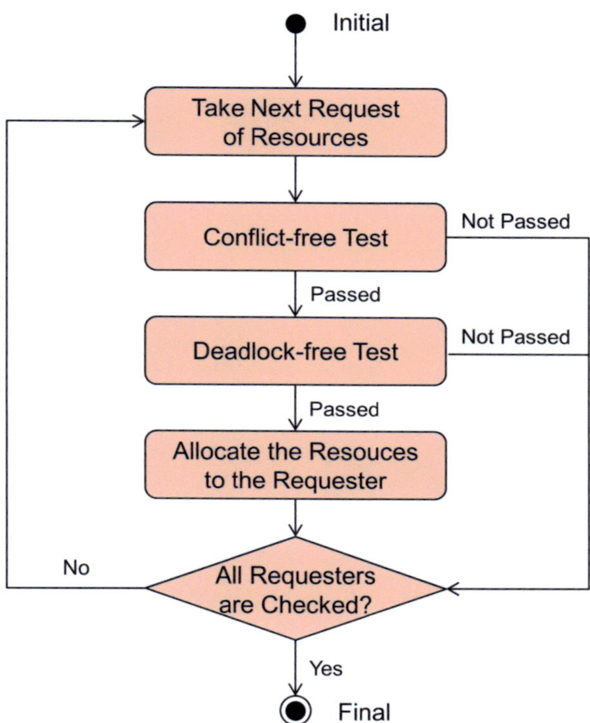


**Figure 3-17: The Process of Resource Allocation (modified from [Cui, 2010])**

### 3.1.2.3    Proceed with Simulation Tasks

After all resource allocations have been processed, the simulation tasks with the state of "Running" will be executed in the current time interval. In a single processing step in one time interval, the information of the three components of the simulation model (i.e. infrastructure resources, simulation performers and simulation tasks) should be updated properly according to different conditions. Similar to the case classification in the process of resource requirement, the situations are also classified into different cases in the process of proceeding with simulation tasks (Figure 3-18). In order to distinguish with the request-cases, a certain Case X defined for proceeding with simulation tasks is named as Proceed-Case X in the following context. Depending on the current characteristics of each simulation task, a matching proceed-case will be designated. The proceed-cases are also divided into two groups according to the ATP systems, and they will be elaborated separately in the following context.

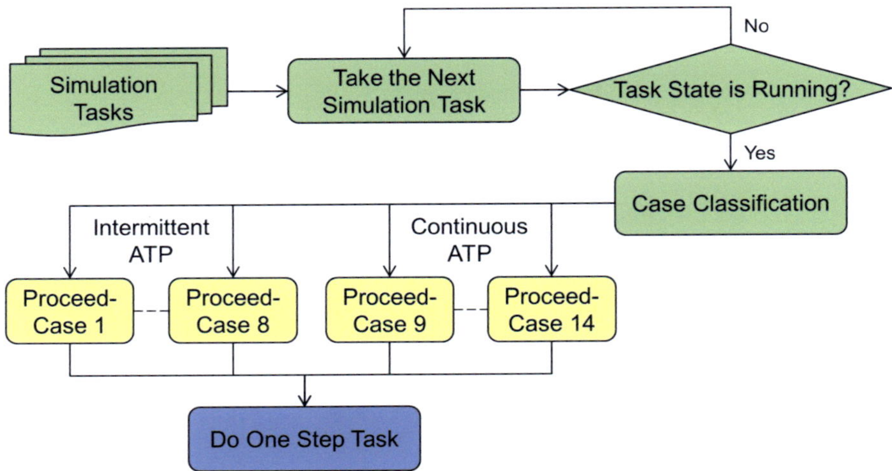


**Figure 3-18: Process of Proceeding with Simulation Tasks (source: [Martin and Liang, 2017])**

In the first group of proceed-cases (Proceed-Case 1 to Proceed-Case 8) the signaling of the concerned train is intermittent ATP. In Proceed-Case 1, the action of the train is "Run" or "Pre-stop", and the current speed of the train is not equal to zero. The train head is located between the entrance signal of the last current block section and the distant signal for the next block section at the beginning of the current time interval, and it is expected to exceed the distant signal at the end of the current time interval (e.g. Train 1 in Figure 3-19). Two subcases are included in Proceed-Case 1. In the first subcase the action of the train is "Run", and the required resources in the next block section are obtained (corresponding to Request-Case 1 of resource requirement). In case the maximum speed limit of the next block section is reduced, a brake application point for speed reduction (e.g. P2 in Figure 3-19) should be calculated and saved in the last current block section. In the second subcase the action of the train is "Pre-stop" or the required resources in the next block section are not obtained (corresponding to Request-Case 1 of resource requirement). A brake application point for stop (e.g. P1 in Figure 3-19) should be determined.

A brake application point in a block section equipped with the intermittent ATP system is used to indicate a specific position for a train, from which the train should start to brake until the train stops in the block section or the head of the train passes the rear signal of the block section. After the train has stopped before or passed the rear signal, the braking application point in the block section will be removed. A brake application point is only necessary when the maximum speed limit of the newly obtained block section is reduced or the resource requirement was rejected. An example is shown in Figure 3-19. If Train 1 (Z1) has obtained the next block section and reduction of maximum speed limit is identified, a brake application point should be determined for Z1. Based on the position of the entrance signal and the maximum speed limit of the next block section, a speed-distance curve for the braking section can be calculated for the train (dashed red line). Based on the current position and speed of the train, a speed-distance curve for the acceleration section and/or constant movement can be calculated (green line). The intersection of these two curves is the brake application point for speed reduction (P2). If the required resources were rejected, the train should stop in front of the rear signal of the last current block section. Correspondingly a new speed-distance curve for the braking section can be calculated (red line) and another brake application point (P1) for stop can be determined.

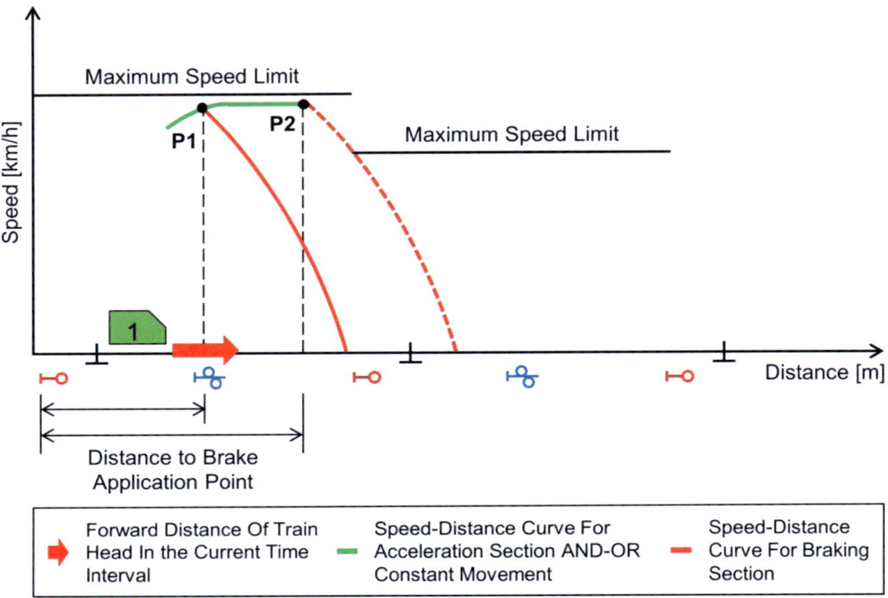



**Figure 3-19: Proceed with Simulation Tasks – Proceed-Case 1 (source: [Martin and Liang, 2017])**

In both of the subcases, the forward distance of the train head should be re-estimated, because the brake application point may influence the movement behavior of the train in the current time interval. The train head position and the speed of the train will be updated in the current time interval. Depending on the new train position, the resources behind the train rear will be released by the train (Figure 3-20). Additionally, if the new position of the train rear has exceeded the rear signal of the first current block section, the first current block section should be removed from the list of current block sections.

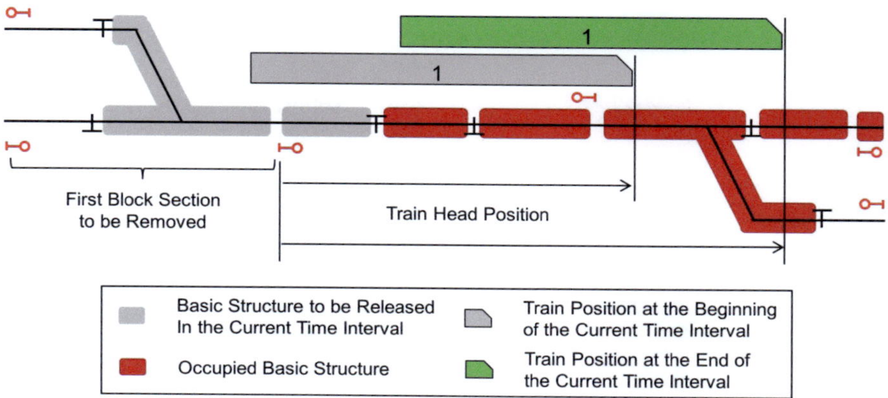


**Figure 3-20: Update of Train Head Positon and Release Basic Structure and Block Section**

Furthermore, if the newly obtained block section is equipped with the continuous ATP system (Figure 3-21), the signaling system of the train should be switched to continuous ATP, and the braking application point (P2) should be removed and a new fixed stopping point (P3) added in case of speed reduction. Once the nominal stopping point[11] of the train exceeds the fixed stopping point for speed reduction, the train should start to brake until the train head enters the next block section.

In addition, if the next block section is a scheduled stop, the action of the train should be changed into "Pre-stop". For a train regulated by intermittent ATP, as long as the train obtained new infrastructure resources, it is necessary to check whether the signaling system needs to be switched and whether the action of the train needs to be changed.

---

[11] During the operation of a train run, based on the attributes of the train at a certain time instance, a braking curve can be calculated, and the end point of the braking curve is considered the nominal stopping point for this train at this moment. With the changes of train attributes (e.g. position and speed), the nominal stopping points change as well. In Section 3.1.2.1 the nominal stopping point is used to detect new resource requirement in the continuous ATP system.

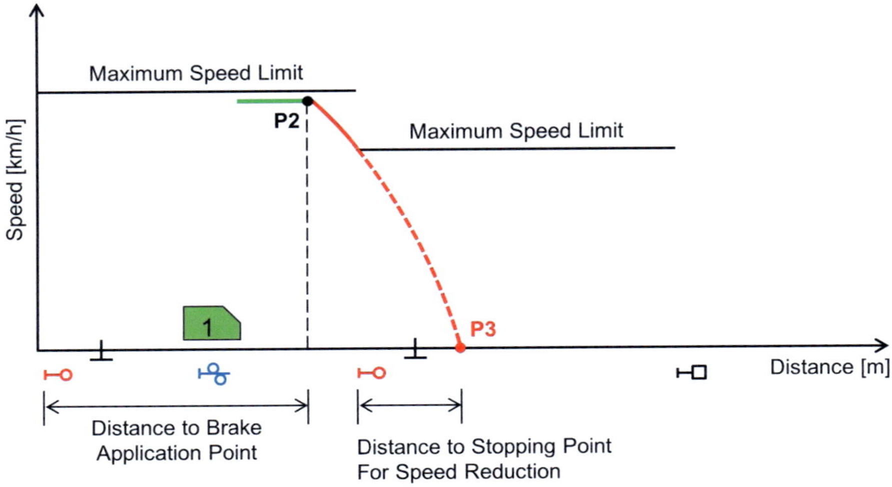


**Figure 3-21: Switching of ATP System – From Intermittent ATP to Continuous ATP (source: [Martin and Liang, 2017])**

Similar to the brake application point for the intermittent ATP system, a stopping point for the continuous ATP system is introduced to support the regulation of train movements in case of maximum speed limit reduction or scheduled or unscheduled stops. Generally, a stopping point is characterized by three attributes: source block section, target block section and the distance between the stopping point and the entrance block marker of the target block section.

In case of maximum speed limit reduction (Figure 3-22), the source block section refers to the one whose maximum speed limit is firstly reduced, and the target block section refers to the one where the stopping point is physically located. Due to the high speed limit on railway line with continuous ATP system, the braking distance may cover more than one block section, so both the source and target block sections are essential for the definition of a stopping point. Once the nominal stopping point of a train exceeded a given stopping point for speed reduction, the train should start to brake until the train head enters the source block section. After the train has entered the source block section at a safe speed (lower than the maximum speed limit of the block section), the stopping point for speed reduction should be removed.

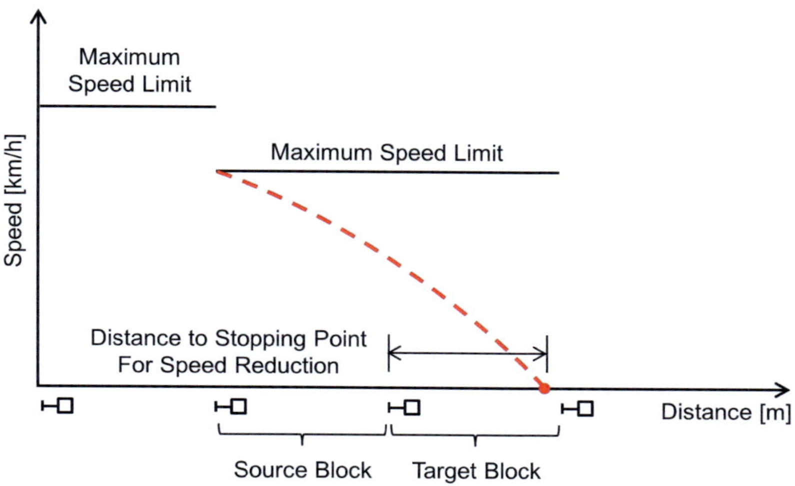


**Figure 3-22: Stopping Point for Maximum Speed Limit Reduction (source: [Martin and Liang, 2017])**

In Proceed-Case 2 the action of the train is "Run" or "Pre-stop", and the current speed of the train is not equal to zero. The train head is located between the distant signal for the next block section and the rear signal of the last current block section, and it is expected to exceed the distant signal for the after-next block section at the end of the current time interval (Figure 3-10). Proceed-Case 2 can be treated in the same way as Proceed-Case 1, the only caveat is that the after-next, instead of the next block section, is required in Request-Case 2 if the action of the train is "Run" (See Request-Case 2 of resource requirement). Accordingly, small details of the algorithm are adjusted, but the procedures involved and the approach are the same. Moreover, the train head will exit the last current block section, so the next block section should be added into the current block section list of the train.

In Proceed-Case 3, the action of the train is "Run", and the current speed of the train is equal to zero (corresponding to Request-Case 3 of resource requirement). If the requirement of the next block section is rejected, nothing is required to be done; otherwise the following procedures should be executed: if the newly obtained block section is a scheduled stop, the action of the train should be changed into "Pre-stop"; if the newly obtained block section is equipped with continuous ATP, the signaling system of the train should also be switched; lastly, the current speed, train head position and current block sections for the train are to be updated and the basic structures behind the rear of the train released, if necessary. These procedures have been discussed in detail, and they will be referred to hereinafter as common procedures.

In Proceed-Case 4, the action of the train is "Turnaround" (corresponding to Request-Case 4 of resource requirement). If the required resources are not obtained, do nothing; otherwise the action of the train should be changed into "Run", and the other procedures to be carried out are depending on the position of train head. Unlike scheduled or unscheduled stops, the relative positional relationship between the train head and the rear signal of the last current block section is irregular in case of turnaround.

If the next block section belongs to the intermittent ATP territory, there are three possible positions of the train head (Figure 3-23). If the train head will not exit the last current block section in the current time interval (Train 1), in addition to the common procedures, the maximum speed limit of the next block section should be checked. A brake application point is necessary in case of speed reduction. If the train head will exit the last current block section, but not exceed the distant signal for the after-next block section (Train 2), only the common procedures should be executed. In case the train head will exceed the distant signal for the after next block section (Train 3), the requirement of the after-next block section will be omitted in the next time interval if train head position moved forward as usual (similar to Request-Case 3 of resource requirement shown in Figure 3-12). Due to the irregularity of train head position, an adjustment of stopping point is not suitable in this case. To ensure logical integrity, a simple and effective solution is implemented: the train head position remains unchanged, and the speed of the train is set to a relatively small value (e.g. 0.001 km/h) in the current time interval. In the next time interval, this proceed-case will be treated as Proceed-Case 2 of proceeding with simulation tasks. The involved train has, therefore, only been delayed for one second (one time interval), which is negligible.

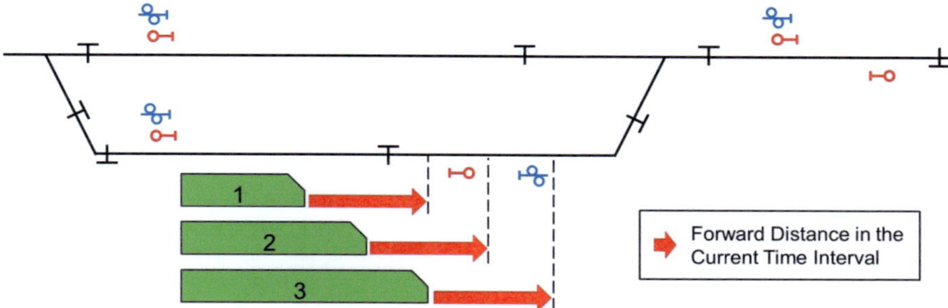


**Figure 3-23: Possible Positions of Train Head In Case of Turnaround**

If the next block section belongs to the continuous ATP territory, there are two possible positions of the train head: in the current time interval the train head position will exit the last current block section (similar to Train 2 in Figure 3-23) or will not exit the last current block section (similar to Train 1 in Figure 3-23). In both cases, the signaling system of the train should be switched to continuous ATP, and the common procedures should be executed. Furthermore, in the second case, the necessity of stopping point for speed reduction should be checked.

In Proceed-Case 5, the action of the train is "Stop". Nothing needs to be done during the dwell time. At the end of the dwell time, if a turnaround task is scheduled, according to the current position of the train rear, the position of the train head must be reset. Depending on the new position of the train, the list of current block sections should be reset. Finally, the action of the train will be changed into "Turnaround". In case of no turnaround task arrangement, it is only necessary to change the action of the train into "Run".

In Proceed-Case 6, Proceed-Case 7 and Proceed-Case 8, the action of the train is "Run" or "Pre-stop" and the current speed of the train is not equal to zero. In Proceed-Case 6, the train head position is located between the distant signal for the next block section and the rear signal of the last current block section, and it will not exit the last current block section (Train 1 in Figure 3-24). Besides the common procedures, if the speed of the train becomes zero at the end of the current time interval, the following procedures should be executed: release the overlap of the last current block section; remove the brake application point in the last current block section; and, if the action of the train is "Pre-stop", change the action into "Stop" and start counting the dwell time. In Proceed-Case 7, the train head position is located between the entrance signal of the last current block section and the distant signal for the next block section, and it will not exceed the distant signal in the current time interval (Train 2 in Figure 3-24). In Proceed-Case 8, the train head is located between the distant signal for the next block section and the rear signal of the last current block section, and it will exceed the rear signal but not the distant signal for the after next block section in the current time interval (Train 3 in Figure 3-24). In both of Proceed-Case 7 and Proceed-Case 8, only the common procedures should be executed.

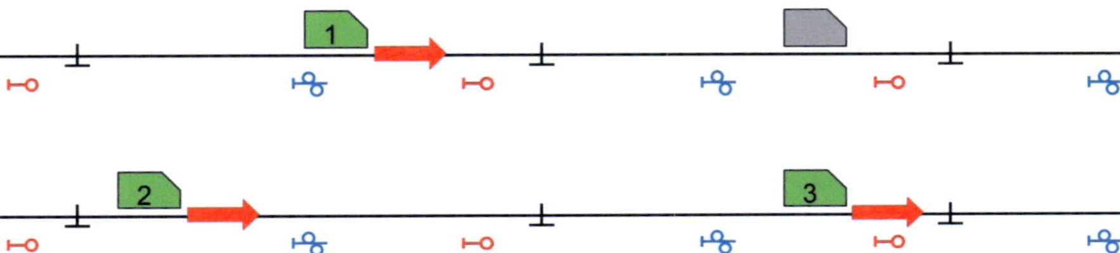

**Figure 3-24: Proceed with Simulation Tasks – Proceed-Case 6, 7, 8**

In the second group of proceed-cases (Proceed-Case 9 to Proceed-Case 14) the signaling of the concerned train is continuous ATP. In Proceed-Case 9, the train speed is not equal to zero and the action of the train is "Run" or "Pre-stop". In addition, the train head will not exit the last current block section or the next block section is still belonging to the continuous ATP territory. Proceed-Case 9 only involves the activities in continuous ATP territory. Depending on the position of the nominal stopping point, two subcases can be elaborated.

In the first subcase, the nominal stopping point of the train is expected to pass a block marker in the current time interval. If the block section beyond the block marker was requested (the action of the train must be "Run") and has been successfully allocated to the train (corresponding to Request-Case 5 of resource requirement illustrated in Figure 3-15), the following procedure should be executed in addition to the common procedures: a stopping point for speed reduction should be determined and the forward distance of the train head should be re-estimated if necessary (Figure 3-22); the action of the train should be changed into "Pre-stop" if the newly obtained block section is a scheduled stop.

If the resource requirement is rejected (the action of the train must be "Run"), a temporary stopping point should be added in front of the entrance block marker of the requested block section (Figure 3-25). The source block section of a temporal stopping point can be set to null, and target block section is the one immediately before the required block section. With the temporary stopping point, the forward distance of the train head in the current time interval will be re-estimated. The lifecycle of a temporary stopping point is only one time interval, because in the next time interval the resource allocation may change. In case that a scheduled stop is to be performed, a new resource requirement is not allowed because the action of the involved train is "Pre-stop". In principle, the movement behavior of the train is the same regardless of whether new resource requirement is not allowed or rejected. Thus, scheduled stops are treated as the same as unscheduled stops. By means of the temporary stopping points in a series of time intervals, the train will be decelerated step by step, and eventually stopped in the block section designated as the scheduled stop.

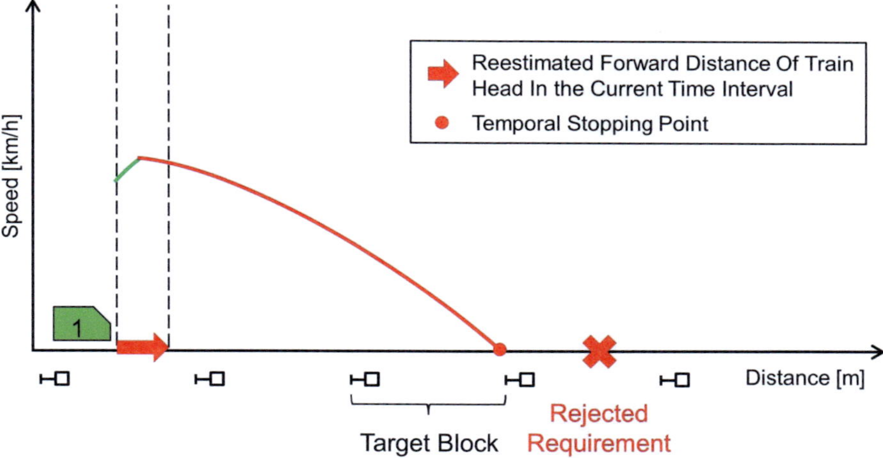


**Figure 3-25: Temporal Stopping Point in Continuous ATP Territory (source: [Martin and Liang, 2017])**

If the resource requirement is rejected or the action of the train is "Pre-stop", based on the re-estimated forward distance of the train, the common procedures can be executed. Moreover, if the speed of the train becomes zero at the end of the current time interval and the action of the train is "Pre-stop", the action should be updated to "Stop".

In the second subcase, the nominal stopping point of the train is not expected to pass a block marker in the current time interval. Therefore, there is absolutely no resource requirement, only the common procedures should be executed.

In Proceed-Case 10, the action of the train is "Run" or "Pre-stop", and the current speed of the train is not equal to zero. The train head will exit the last current block section, and the next block section belongs to intermittent ATP territory. Furthermore, the train head will pass the distant signal for the after next block section (corresponding to Request-Case 6 of resource requirement as shown in Figure 3-16). Firstly, this proceed-case can be treated as the same as the first subcase of Proceed-Case 9 and the same procedures should be carried out, since the signaling system of the train is still continuous ATP. The updated position of the train head is located on a new block section belonging to the intermittent ATP territory. Correspondingly, the signaling system of the train should be switched, and the train movement in the next time interval will be regulated by intermittent ATP system. As shown in Figure 3-26, if a stopping point for speed reduction (P2) or a temporal stopping point (P1) was created in the current time interval, it should be replaced by a brake application point (P3 or P4). To estimate forward distance in one time interval, only brake application points are considered for trains under the mode of intermittent ATP system, and only stopping points are taken into account for trains under the mode of continuous ATP system.

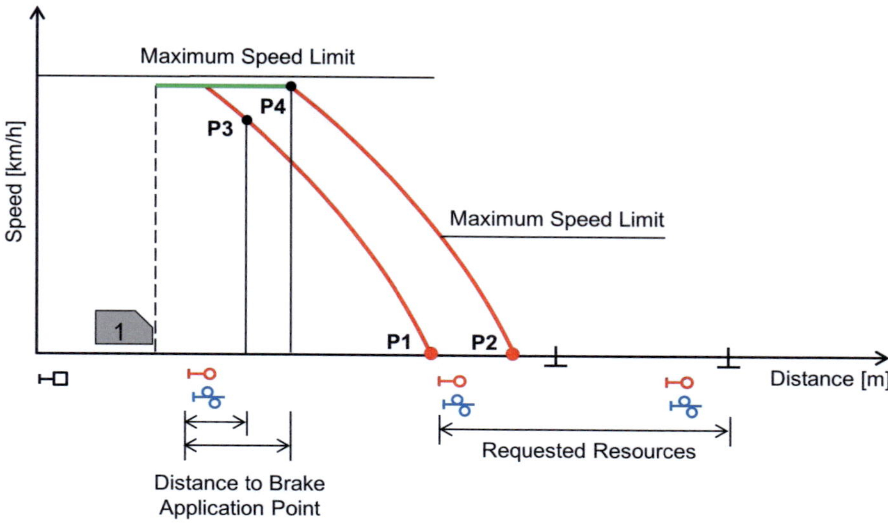


**Figure 3-26: Switching of ATP System – From Continuous ATP to Intermittent ATP**

From the examples shown in Figure 3-21 and Figure 3-26, it can be seen that the boundary conditions for switching the ATP system of a train from intermittent mode to continuous mode and from continuous mode to intermittent mode are defined differently in the simulation model. For a train under the mode of intermittent ATP, once the train has obtained a block section equipped with continuous ATP, the signaling system of the train must be switched in order to ensure the seamless connection of the two ATP systems. Before the train head has physically left the intermittent ATP territory, new resource requirements can occur in accordance with the resource requirement principle of the continuous ATP system (Figure 3-27). If the signaling system would not be switched until the train head leaves the intermittent ATP territory, the block section that ought to be requested (marked in Figure 3-27) would be omitted, which would lead to a logic error of the simulation model.

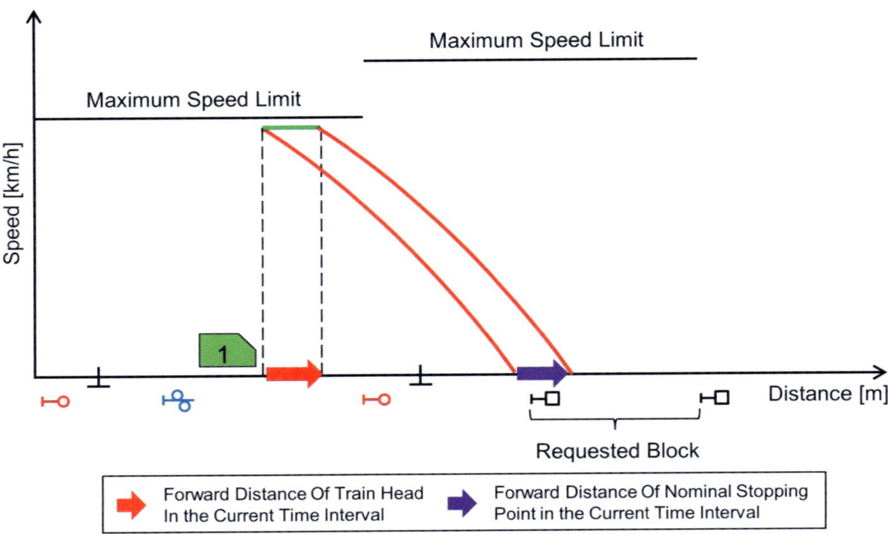


**Figure 3-27: Resource Requirement in Transition Zone between Intermittent and Continuous Territories**

For a train under the mode of continuous ATP, only if the train head has already physically entered a block section equipped with intermittent ATP can the signaling system of the train be switched. An example is shown in Figure 3-28, wherein a train decelerates in the current time interval because of speed reduction. If the signaling system of the train had been switched into intermittent ATP, on one hand, the stopping point for speed reduction would not be detected, which would cause loss of control of train movement (the train would accelerate in the current time interval); on the other hand, the stopping point for speed reduction cannot be replaced by a brake application point, because brake application points are only allowed to be created in intermittent ATP territory. Therefore, in the design of simulation model with multiple signaling systems, special attention should be paid to switching signaling system in transition zones between different signaling system territories.

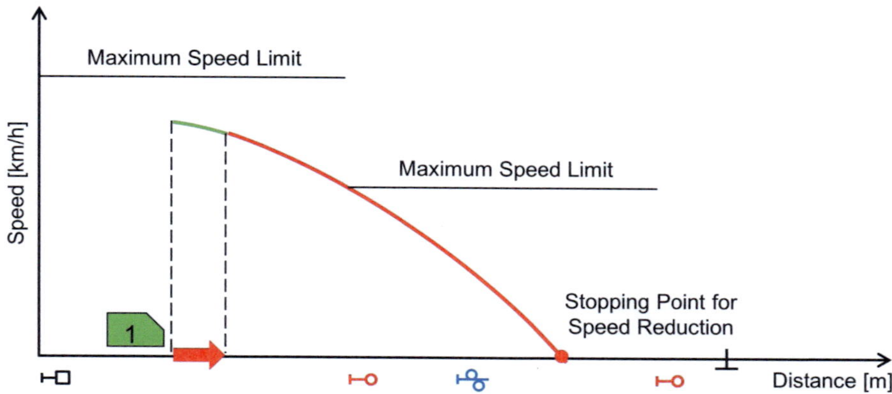


**Figure 3-28: Train Speed Control in Transition Zone between Continuous and Intermittent Territories**

In Proceed-Case 11, the current speed of the train is equal to zero, and the action of the train is "Run" (corresponding to Request-Case 7 of resource requirement). If the requested resources are obtained, the common procedures are to be executed; otherwise, nothing is to be done. In Proceed-Case 12, the action of the train is "Turnaround" (corresponding to Request-Case 8 of resource requirement). In principle, the procedures involved in this proceed-case and Proceed-Case 4 of proceeding with simulation tasks are the same, special attention should be paid to the different manners of switching signaling system as explained above. In Proceed-Case 13, the action of the train is "Stop". This proceed-case can be treated as the same as Proceed-Case 5 of proceeding with simulation tasks. In Proceed-Case 14, the current speed of the train is not equal to zero and the action of the train is "Run" or "Pre-stop". The train head will enter the next block section belonging to intermittent ATP territory, but not pass the distant signal for the after next block section (Figure 3-29). The common procedures should be carried out, and the signaling system of the train should be switched.

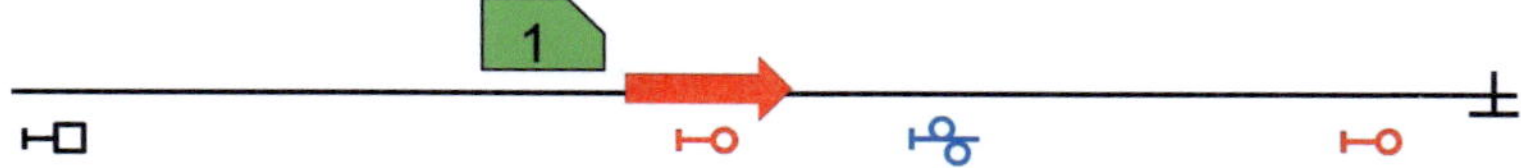


**Figure 3-29: Proceed with Simulation Tasks – Proceed-Case 14**

## 3.2 Meso- and Macroscopic Simulation Model

In order to ensure that the simulation can be concurrently carried out on the microscopic, mesoscopic and macroscopic levels, the meso- and macroscopic simulation models also follow the same workflow of synchronous simulation used by the microscopic model. In the meso- and macroscopic simulation models, the attributes of infrastructure resources, simulation performers and simulation tasks were simplified to different extents to improve the computational performance of the simulation model, and the procedures included in the workflow - request resources, allocate resources and proceed with simulation tasks - were also accordingly modified. In order to elaborate the meso- and macroscopic model clearly, the hierarchy of infrastructure has to be defined primarily (Section 3.2.1).

### 3.2.1 Infrastructure Hierarchy

In general, the infrastructure classification used in this approach complies with the principle of macroscopic infrastructure modelling outlined in [Radtke, 2014]. Even through the principle is originally designed for macroscopic models, it is expanded to apply to multi-scale models in this approach. Three types of nodes are defined on the basis of their topological characteristics as follows: loop nodes, junction nodes and open track sections. Within a loop node two further categories are defined: loop track and loop non track. The classification diagram for infrastructure hierarchy is shown in Figure 3-30. As described in Section 3.1.1.1, basic structures are basic elements constituting an entire network. They are also basic elements in this infrastructure classification. A basic structure belongs only to one node (i.e. a loop node or a junction node or an open track section), and the corresponding node will be recorded as an attribute, called a node attribute, in the basic structure instance. A block section also has a node attribute, which is identical to the node attribute of the last basic structure located at the end of it.

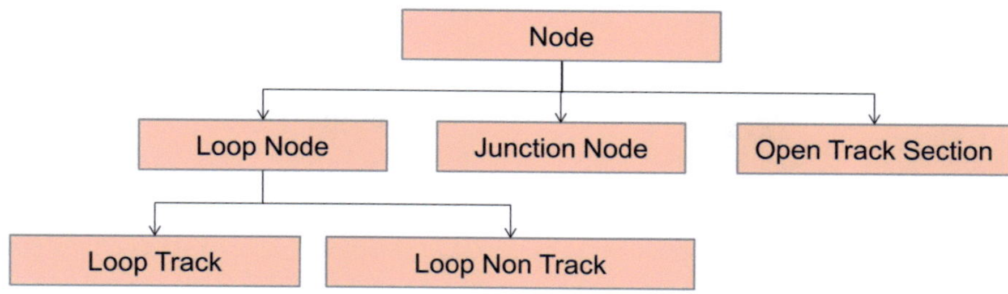

Figure 3-30: Classification Diagram for Infrastructure Hierarchy (source: [Martin and Liang, 2017])

A loop node is defined as an operational site where a train can overtake or pass another train by taking a different operational route within the operational site. If entry signals exist around a loop node, they are taken as boundaries of the loop node. Otherwise the first signal outside of the loop node is used as a boundary. Loop tracks are the tracks that can be designated as scheduled or unscheduled stops for trains in a loop node. The other components of the loop node (except loop tracks) are generally defined as loop non track. A junction node includes at least one junction-type resource, but it is impossible to arrange an overtaking or passing operation inside the junction node. Apart from loop nodes and junction nodes, the other parts of the network are referred as open track sections. The virtual block section is a special type of open track section, which consists only of free resources. As an example, the different types of nodes on a reference infrastructure network are shown in Figure 3-31.

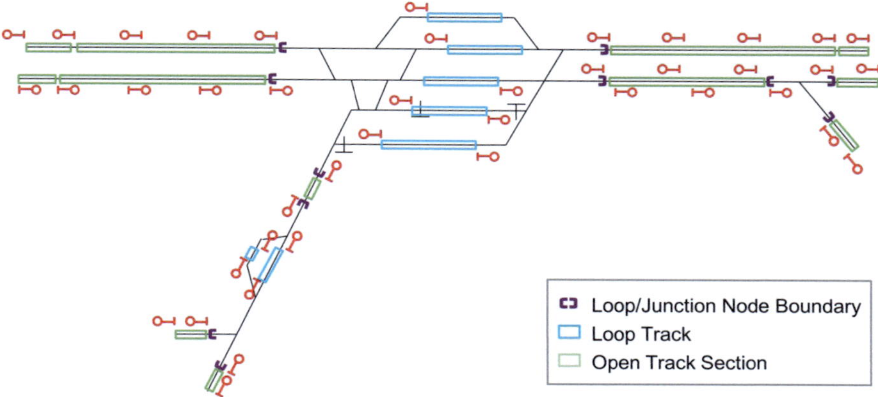


Figure 3-31: An Example of Different Infrastructure Nodes

## 3.2.2    Mesoscopic Simulation Model

Due to the state of the art a universal accepted mesoscopic simulation model for railway operation does not exist, at present. The design of the mesoscopic simulation model varies, as well, in different application contexts. In the mesoscopic simulation model developed in [Marinov and Viegas, 2011], the investigated network is separated into interconnected components such as railway stations, open track sections between stations (double and single tracks), shunting yards and so on. The components are modeled in a queuing system to evaluate the influences of freight train operations in a railway network. The service pattern of a server is described by a concrete distribution obtained from actual data and statistical analysis. For instance, the running time on an open track section is set as a deterministic value with an additional random component. The mesoscopic model developed in [Fabris et al., 2014] is used to generate railway timetables. In this model, running time is calculated by solving train motion equations, which is similar to the microscopic model. The details of description of loop tracks and open track sections approximate microscopic level, while the interlocking areas at each side of station tracks are simplified to matrices of train routes and their compatibility. The same method is also used by [Corman et al., 2009] to describe interlocking areas in his microscopic model, in which running times on block sections are given as deterministic values. The mesoscopic infrastructure model defined in [Radtke, 2014] refers to the mesoscopic open track sections and stations automatically generated based on a given macroscopic network. Open track sections are composed of standard block sections, and stations are modeled with individual station tracks and possible train routes. The mesoscopic infrastructure model standardized by [UIC, 2016] focuses on the description of open track sections between the operational points of the network. An operational point is defined as "any location for train service operations, where train services can begin and end or change route, and where passenger or freight services are provided" in [UIC, 2014]. For instance, stations and shunting yards belong to operational points. The operational points in a network are abstracted into points, while the details of open track sections between them are kept on an almost microscopic level (e.g. aggregation example shown in Figure 3-32).

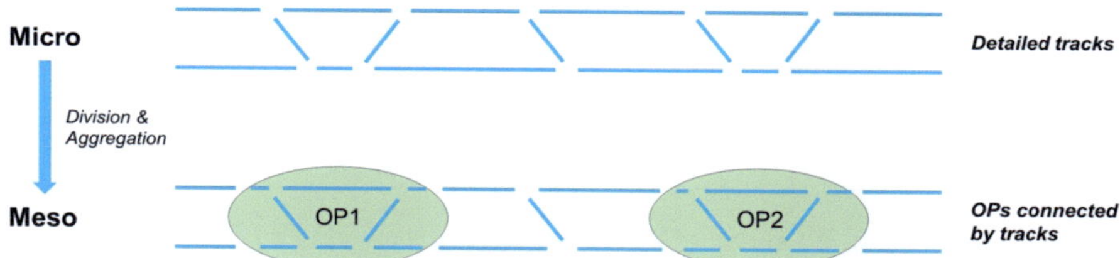


**Figure 3-32: Aggregation Example [UIC, 2016]**

The above-mentioned literature reveals that even though the levels of detail of the mesoscopic models differ, the methods of simplification employed are similar:

–      Mostly constant running time is used instead of exact running time calculation

–      Routes through area of loop non track (interlocking areas) can be simplified to different extents

–      Due to the relatively simple topology of open track sections and loop tracks (stations tracks), their potential of further simplification on the mesoscopic level is limited.

Following these common approaches, a comprehensive mesoscopic model characterized by continuous scaling was developed in this dissertation. The mesoscopic model can achieve any level of detail through the stepwise simplification of the topology of an infrastructure network. Importantly, the mesoscopic model does not contradict but represents a good supplement to the standard published by UIC (International Union of Railways).

### 3.2.2.1      The Components and Workflow on Mesoscopic Level

On a mesoscopic level, constant maximum average running time is implemented instead of accurate running time calculations. So only one behavior section – constant movement – is considered in the estimation of forward distance covered in one time interval (acceleration section and braking section are excluded). The maximum allowed speed of a train at a certain time instance is the minimum value between the maximum speed of the train and the maximum speeds of current block sections of the train (see Formula (I-11) in Appendix I). The expected forward distance of the head of a train can be simply calculated as:

$$S = v_{max} \cdot t \qquad\qquad (3\text{-}11)$$

Notation used:

$S$:    Expected forward distance of the head of a train in the current time interval

$v_{max}$: Maximum allowed speed of a train at a certain time instant

$t$:    Length of a time interval

On the most detailed mesoscopic level, the infrastructure network topology is kept the same as on the microscopic level. On more abstracted mesoscopic levels, the topology could be simplified stepwise by the aggregation of basic structures to different extents. The simplification procedure will be elaborated separately in Section 3.2.2.2. One basic principle of the mesoscopic infrastructure model is that block sections will continue to be used in the workflow on the mesoscopic level.

In the microscopic model, the distant signal and the relative position of nominal stopping points and block markers are used to trigger new resource requirements. If new resource requirements are not obtained, the concerned trains should start braking timely and halt before stop signals accurately. However, owing to the employment of constant running time in the mesoscopic model, it is meaningless to continue to use the distant signal and the relative position of nominal stopping points and block markers as indicators of resource requirements. Therefore, two simple request-cases are summarized to detect new resource requirements in the mesoscopic model (unless noted otherwise, the case numbers used in this section refer to these two request- cases defined for the mesoscopic model):

Request-Case 1: if the action of a train is "Run" and the train head is expected to exceed the rear signal or block marker of the last current block section, the basic structures included in the next block section will be requested (Figure 3-33).

Request-Case 2: if the action of a train is "Turnaround", the basic structures beyond the train head in the last current block section and the ones in the next block sections will be requested (the same as Request-Case 4 and Request-Case 8 of requesting resources in the microscopic model).

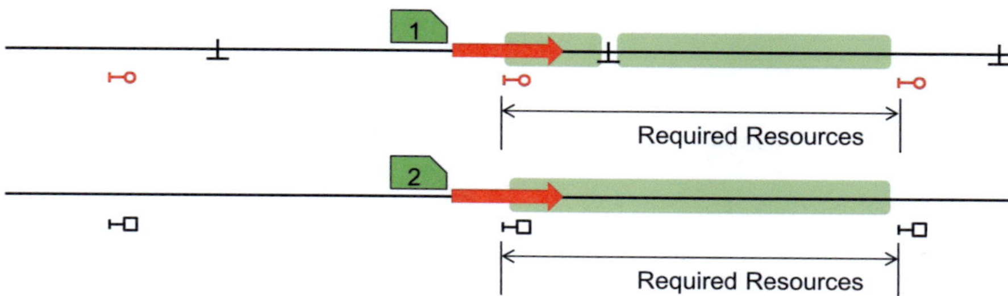

**Figure 3-33: Request Resources – Request-Case 1 in Mesoscopic Model (source: [Martin and Liang, 2017])**

In Figure 3-33, it can be seen that overlaps still exist on the most detailed mesoscopic level, but they will not be requested together with their corresponding block sections. On more abstracted mesoscopic levels, overlaps may disappear on account of aggregation of basic structures, and therefore, it is convenient to implement a consistent logic of resource requirement (without consideration of overlaps) in the whole mesoscopic model. With the above-mentioned simplifications, the processes of requesting resources in intermittent ATP and continuous ATP territory become identical, which can be described by Request-Case 1 and Request-Case 2 defined previously. Accordingly, the signaling system of trains is no longer considered in the mesoscopic model.

No matter the resource requirement occurred in the micro-, or meso-, or macroscopic simulation model, all resource requirements should be processed in a centralized manner as described in Section 3.1.2.2.

With regard to proceeding with simulation tasks, only the simulation tasks with the state of "Running" will be processed in the current time interval. Depending on the action and the position of a train, four proceed-cases are classified to cover all possible situations (unless noted otherwise, the case numbers used in this section refer to these four proceed-cases defined for the mesoscopic model).

In Proceed-Case 1, the action of the train is "Run" or "Pre-stop" and the train head is expected to exit the last current block section in the current time interval. There are two subcases included in Proceed-Case 1. In the first subcase, the action of the train is "Run" and requested resources are obtained (corresponding to Request-Case 1 of requesting resources). The common procedures[12] should be executed, which include:

- The action of the train should be changed to "Pre-stop" if the new obtained block section is a scheduled stop;

- The train head position and current speed should be updated (in case of speed reduction the forward distance should be recalculated as shown in Figure 3-34);

- Update the list of current block sections for the train, and release the basic structures behind the new position of the train rear.

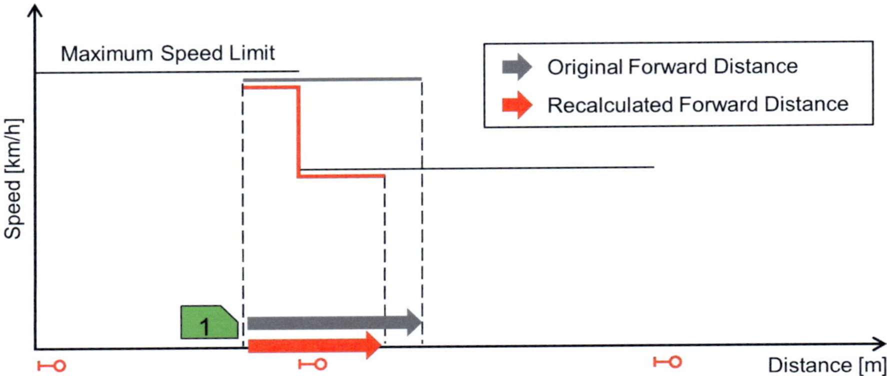


**Figure 3-34: Update Train Head Position in the Mesoscopic Model (I) (source: [Martin and Liang, 2017])**

In the second subcase the next block section is not obtained, the train head will be updated to the position where the rear signal of the last current block section locates (Figure 3-35). Moreover, common procedures should be executed. When the action of the train is "Pre-stop", it should be changed to "Stop" and counting of dwell time should be started.

---

[12] The common procedures are also used in the process of proceeding with simulation tasks in the microscopic model.

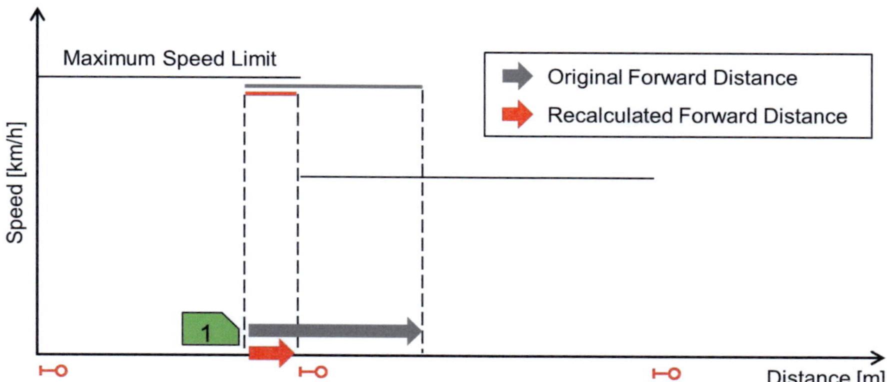


**Figure 3-35: Update Train Head Position in the Mesoscopic Model (II)**

In Proceed-Case 2, the action of the train is "Run" or "Pre-stop" and the train head position will not exceed the last current block section in the current time interval, in which case only the final two common procedures should be carried out.

In Proceed-Case 3, the action of the train is "Turnaround" (corresponding to Request-Case 2 of requesting resources). If the requested resources are not obtained, nothing is to be done; otherwise, the action of the train should be updated to "Run", and the other procedures to be executed are dependent upon the expected position of the train head at the end of the current time interval:

- If expected position exceeds the last current block section, the same procedures for the first subcase of Proceed-Case 1 will be executed,

- If the expected position is still within the last current block section, the same procedures for Proceed-Case 2 will be executed.

In Proceed-Case 4, the action of the train is "Stop", which is the same as Proceed-Case 5 and Proceed-Case 13 of proceeding with simulation tasks in the microscopic model (for details see Section 3.1.2.3).

### 3.2.2.2 Simplification of the Infrastructure Network Topology

For the more abstracted mesoscale models, two or more adjacent basic structures will be combined into a larger mesoscale occupation unit. Because block sections will continue to be used on the mesoscopic level, aggregation of basic structures should not violate the division of block sections. This means only if none of the connection node(s) between two adjacent basic structures is a main signal or block marker, the two basic structures are allowed to be aggregated.

Basic structures are composed of edges, so the aggregation of two adjacent basic structures is realized through a combination of adjacent edges in the same direction. A schematic diagram of an infrastructure is shown in Figure 3-36 as an example: two basic structures (BS_L and BS_R) are to be aggregated, and one connection node (Node 5) exists in between. Each edge from BS_R is physically connected with two edges from BS_L (red dashed line in the subgraph "Basic Structure"). To maintain these connection relationships after aggregation, each edge from BS_R should be duplicated once (subgraph "Edge Duplication"). Thus, on both sides of Node 5, the amount of edges are the same, and eventually these edges can be combined into two pairs of new edges.

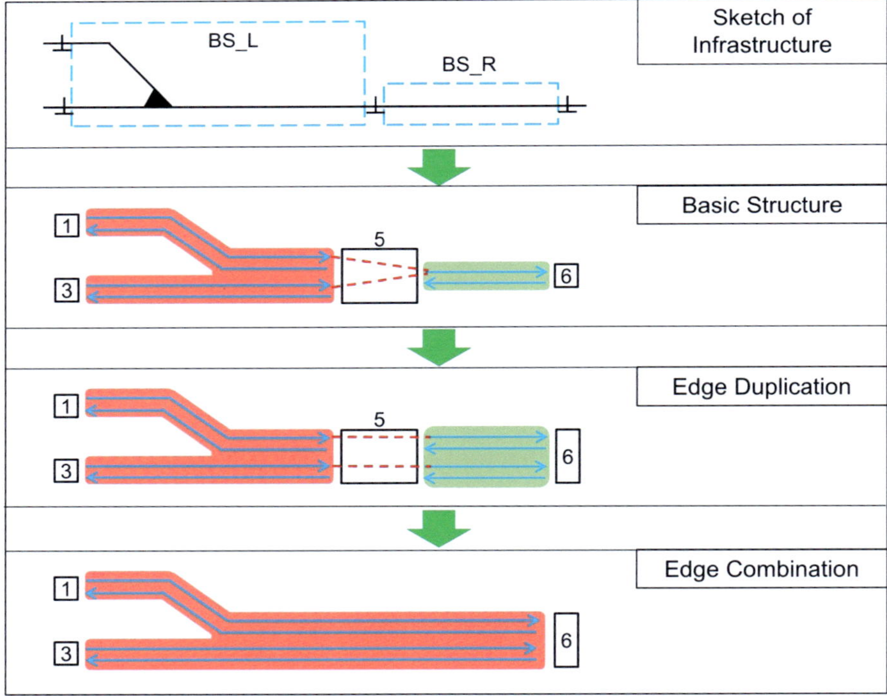


**Figure 3-36: Edge Duplication and Combination for Aggregation of Basic Structures**

It can be seen from the example above that edge duplication may have to be carried out as the preparation of basic structure aggregation. Generally speaking, supposing that there exist m pairs of edges in one basic structure (BS1) and n pairs of edges in another basic structure (BS2), using one connection node as their endpoints; each edge from BS1 should be duplicated (n-1) times, and each edge from BS2 should be duplicated (m-1) times; finally, these edges will be combined into (m*n) pairs of new edges.

The length of a new edge is the sum of the lengths of the included edges, and the new permissible speed of the new edge is the average speed limit considering the proportionate length of the included edges. The new edges and basic structure will be renamed with new IDs. Accordingly, the included edges and corresponding basic structures of the involved block sections should also be updated.

The throughput[13] of a basic structure is one train only, but the throughput of a larger mesoscale occupation unit may vary under different operation situations. Thus, the aggregation accuracy of each possible throughput will be calculated, and the throughput with the highest aggregation accuracy will be chosen as the representative throughput of the larger mesoscale occupation unit. Thus each mesoscale occupation unit has two important attributes: representative throughput and aggregation accuracy. The calculation method of these two attributes will be elaborated in Section 4.2.

### 3.2.3    Macroscopic Simulation Model

Compared to microscopic and mesoscopic models, the macroscopic model is capable of covering a large investigated area with a huge amount of train movements due to its high level of abstraction. According to the infrastructure hierarchy defined in Section 3.2.1, an infrastructure network can be decomposed into three different types of nodes: loop node, junction node and open track section. On microscopic and mesoscopic levels, a node refers to a certain area composed of basic structures or occupation units and block sections. On the macroscopic level, a loop node or a junction node will be abstracted into a vertex, and an open track section will be abstract-

---

[13] The number of trains that can simultaneously occupy the mesoscale occupation unit is defined as its throughput.

ed into a link [Radtke, 2014] [Cui, 2010] [Cui and Martin, 2011]. To keep the terminology consistent, the terms – loop node, junction node and open track section – will to be used continue in the macroscopic model, even though their manifestation (level of detail) has been changed.

In the microscale and mesoscale model, trains are spatially separated by block sections. In the macroscale model, due to the lack of detailed information on block sections, trains have to be temporally separated.

On open track sections, trains can be separated by minimum line headways. The detailed calculation method of minimum line headways can be found in [Pachl, 2014]. Similar to the macroscopic model developed in [Cui, 2010], two kinds of minimum line headways including arrive-arrive headway (denoted by AA) and depart-arrive headway (denoted by DA) are used herein (Figure 3-37). Arrive-arrive headway is used to separate two trains with successive movements that arrive sequentially onto the same open track section. Depart-arrive headway is used to separate two trains with opposite movement, which is the minimum time interval between the departure of the first train from and the arrival of the second train onto the same open track section (the names of headways are modified in order to better illustrate the macroscale model developed in this approach, and original definitions of headways (see [Pachl, 2014]).

Besides minimum line headways, the movement authority onto an open track section is also restricted by the throughput of the open track section. Similar to the throughput of mesoscopic occupation unit, the maximum number of trains that can occupy a macroscopic node simultaneously (i.e. open track section, loop node and junction node) is also defined as its throughput. For a unidirectional open track section the throughput is the number of its included block sections. For a bidirectional open track section, the throughput for each direction should be defined with the same method. When the throughput of an open track section is consumed (the number of current occupiers is equal to the throughput), new request of the open track section cannot be allocated.

On an open track section, overtaking is not possible, so the train sequence is regulated based on the FIFO principle (first-in-first-out).

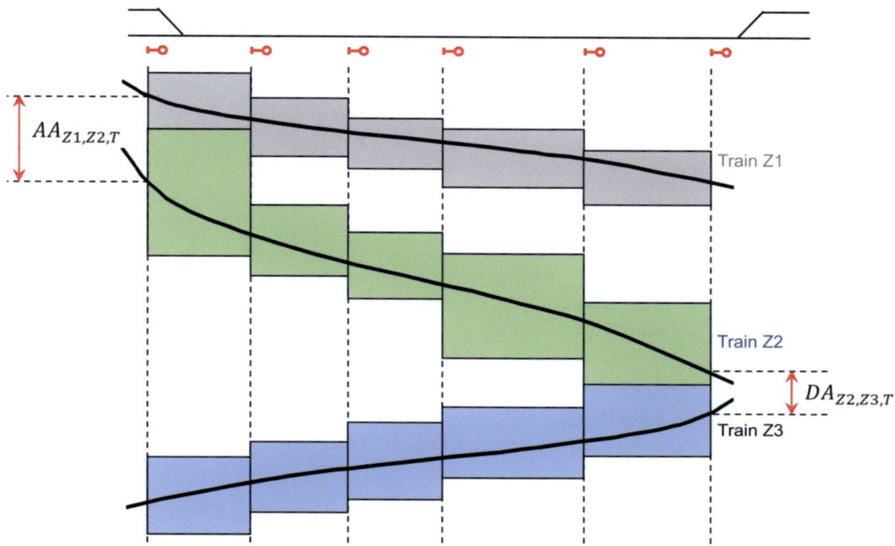


**Figure 3-37: Minimum Line Headway on Open Track Section T (modified from [PT1, 2016] and [Cui, 2010])**

In loop nodes, conflicts between station routes are not modelled on the macroscopic level due to the high level of abstraction. The movement authority into a loop node is only restricted by the throughput of the loop node. As long as the throughput is not yet consumed, the request of the loop node can be directly authorized. In general, there are three methods to describe a loop node on the macroscopic level, and their levels of details are different. For the first description method, the throughput of a loop node is assumed to be infinite. For the second method, throughput will be determined based on track groups in the macroscale model. A track group is defined as a set of loop tracks with identical inbound and outbound open track sections, and the throughput of the track group is the number of the included tracks. To deal with the request of a loop node from a certain train, only the throughput of the related track groups will be checked. For the third description method, the information on the included loop tracks and the connection relationship between the loop tracks and open track sections are stored as shown in Figure 3-38. The throughput of a loop track is set to one. It can be seen that only in the last one the information of all loop tracks are maintained, which is important for dispatching tasks (see Chapter 6). In order to keep the consistency of the dispatching optimization algorithm on different description levels, the third description method is used in this dissertation. Furthermore, in loop nodes, overtaking is allowed, so the train sequences are not restricted.

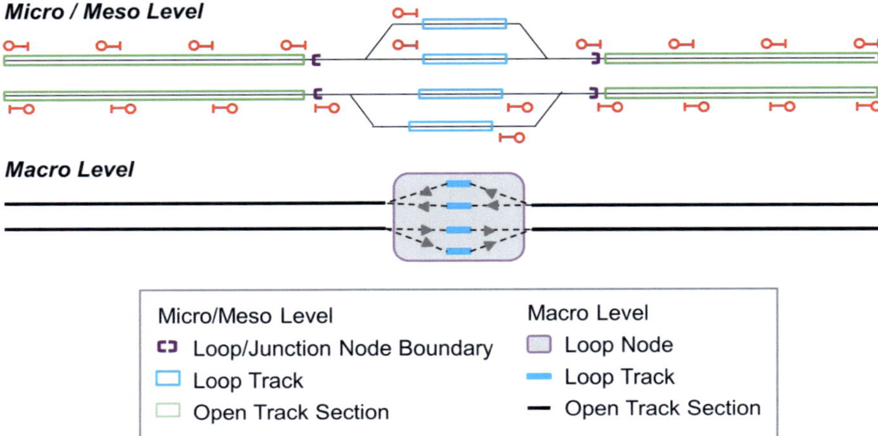


**Figure 3-38: An Exemplary Macroscopic Infrastructure Network**

Similar to loop nodes, the movement authority into a junction node is also restricted by throughput. However, in junction nodes the relative sequence of the trains coming from the same open track section is not allowed to be changed, because overtaking is not possible (e.g. Z1 and Z2 in Figure 3-39). Thus, the train sequence is also regulated according to the FIFO principle in junction node. Because train sequences in loop nodes are not dispatching related variables in this model, the trains attempting to go into a loop node follows the FCFS (First Come First Serve) principle (e.g. Z1, Z2 and Z3 in Figure 3-39).

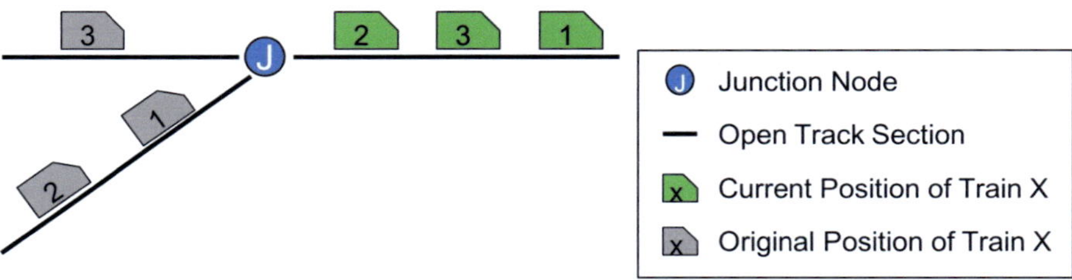


**Figure 3-39: Maintain Train Sequence Consistency in Junction Node**

As described above, trains are temporally separated on the macroscopic level, so only the departing/passing time on each macroscopic node is relevant. Compared to the microscale and mesoscale models, in which the resource requirement is triggered by the position of a train, in the macroscale model the resource requirement is triggered by time. Furthermore, the macroscale model cannot depict the position of a train as exactly as the microscale and mesoscale models, so in the macroscale model the current position of a train refers to its current macroscopic node. Accordingly, the detailed path of a train (see Section 3.1.1.1) is abstracted into a macro-path. The definition in [Cui, 2010] is used herein: a macro-path is a list of macroscopic nodes ordered based on the given timetable. In the simulation process, the position of a train will be updated from node to node along its macro-path.

No matter which kind of macroscopic node the current position of a train is, the prerequisite of resource requirement is that the scheduled operation on the current node should be completed, which can be checked by:

$$T_{now} \geq TB_{i[N]-1,Z_j} + TI_{i[N],Z_j} \tag{3-12}$$

Notation used (the notations are the same or similar to those used in [Cui, 2010]):

$i[N]$:      Index of the macroscopic node N in the macro-path of a corresponding train

$Z_j$:      Train $Z_j$

$T_{now}$:      Current execution time in the simulation model

$TB_{i[N]-1,Z_j}$:      Departing/passing time for train $Z_j$ in the $(i[N] - 1)^{th}$ node of its macro-path

$TI_{i[N],Z_j}$:      Scheduled operation time for train $Z_j$ in the $(i[N])^{th}$ node of its macro-path

The departing/passing time from the last node ($TB_{i[N]-1,Z_j}$) implies the arrival time onto the current node, and the scheduled operation time on the current node ($TI_{i[N],Z_j}$) includes both scheduled running time and scheduled dwell time. The operation time on each macroscopic node should be determined in advance with the assistance of the microscale model. If the constraint of operation time is not yet satisfied for a train at a certain time instance ($T_{now}$), new resource requirement is not allowed. If the scheduled operation is completed, other constraints may also have to be checked according to the current position of the train as follows.

- In loop nodes, the train is allowed to request the next resource, since no other constraints exist.
- In junction nodes or on open track sections, due to the FIFO principle, only if a train is the first occupier of the current macroscopic node will the next resource be requested; otherwise, new resource requirement is not allowed.

In the process of resource allocation, a new resource requirement should also pass the conflict-free and deadlock-free tests. The conflict-free test concerns the constraints of headways and throughput. In case the requested resource is a loop node or junction node and the throughout is not yet consumed, the resource requirement can be directly labeled as conflict-free. In case the requested resource is an open track section, besides the constraint of throughput, the headway times also have to be respected. As described above, there are two kinds of headways (i.e. AA and DA), so, firstly, the running directions of the requester (denoted by $Z_j$) and its immediately previous train (denoted by $Z_{Prev}$) on the open track section (denoted by $T$) should be compared. For successive movements, the constraint of headway is:

$$T_{now} \geq TB_{i[T]-1,Z_{Prev}} + AA_{Z_{Prev},Z_j,T} \tag{3-13}$$

For opposite movements, the constraint of headway is:

$$T_{now} \geq TB_{i[T],Z_{Prev}} + DA_{Z_{Prev},Z_j,T} \tag{3-14}$$

Notation used:

$DA_{Z_{Prev},Z_j,T}$:　　Depart-arrive headway between the train $Z_j$ and $Z_{Prev}$ on open track section $T$

$AA_{Z_{Prev},Z_j,T}$:　　Arrive-arrive headway between the train $Z_j$ and $Z_{Prev}$ on open track section $T$

Before a simulation starts, all departing/passing times should be set to positive infinite. If Formula (3-13) is taken as an example: only after the train $Z_{Prev}$ has physically departed from the resource before T will $TB_{i[T]-1, Z_{Prev}}$ be updated to the actual departing/passing time. It is possible that T has been allocated to $Z_{Prev}$ by dispatching even through $Z_{Prev}$ has not physically departed from the resource before T. In this case, $TB_{i[T]-1, Z_{Prev}}$ is still positive infinite, which results in that $Z_j$ is not allowed to enter T, so $Z_{Prev}$ refers to the last occupier of T.

The resource requirements having passed the conflict-free and deadlock-free tests will be allocated to the corresponding requesters. In the process of proceeding with simulation tasks, if a train obtains new resources in the current time interval, the following procedures should be carried out:

- Remove the train from the occupier list of the current node
- Add the train into the occupier list of the next node
- Record the departing/passing time on the current node
- Update the current position of the train to its next node

If no resource requirement occurred or requested resources were not successfully allocated, do nothing

# 4 Assessment Method for the Multi-scale Model

To balance the computational complexity and simulation accuracy, the multi-scale mode developed in Chapter 3 could be implemented when working with a large investigation area. This means that some infrastructure nodes within the investigation area are simulated on the microscopic level, while others are simulated on the mesoscopic and macroscopic levels. The appropriate description level of a specific infrastructure node is determined by its significance value. The infrastructure nodes with lower significance values will be abstracted and simulated on the mesoscopic or macroscopic level; those with higher significance values will be kept on the microscopic level. In order to determine the significant value of each infrastructure, an assessment method for the multi-scale model was developed within in the DFG project [Martin and Liang, 2017]. The significance value integrates two indicators: the relevance to conflicts and the aggregation accuracy. The relevance to conflicts is positively while the aggregation accuracy is negatively correlated to the significance value. The relevance to conflicts should be on-line determined, while the aggregation accuracy should be off-line prepared in advance. The details of the assessment method will be elaborated in this chapter.

## 4.1 Calculation method of Significance Value

Relevance to conflicts can be determined by means of delay propagation models. Infrastructure nodes located in the propagation scopes of conflicts have a higher degree of correlation than the ones outside the propagation scopes. Delay propagation in different scopes has been studied in some research projects: delay propagation in stations was investigated in [Yuan, 2006], and delay propagation in large scopes was researched in [Goverde, 2010] and [Siefer and Radtke, 2006]. For practical applications, a simplified method was developed based on the multi-scale simulation model to assess propagation scopes instead of accurate delay propagation models.

When conflicts occur, or potential conflicts are detected at a certain instance in time, the traffic condition in the prediction period (e.g. the next hour) can be simulated with the simulation model[14]. In case of conflicts, the hindered train will get a knock-on de-

---

[14] The microscale or mesoscale model is recommended for traffic condition prediction, because these two models can describe the interactions between trains in detail. Due to the high level of abstraction

lay on a certain block section, and the delay will accompany the train on the further block sections along its path until the delay is eliminated by recovery times or by the termination of the train run. As shown in Figure 4-1, a train Z2 on the block section B0 was hindered by another train Z1 on the block section B1, and Z2 obtained a knock-on delay $tw_{2,0}$ of two minutes on B0. The knock-on delay resulted in delays of Z2 $td_{2,i}$ on the further block sections along its path. The delay is recovered stepwise and totally eliminated on B4. In this example, the delay propagation scope covers the area from B0 to B4.

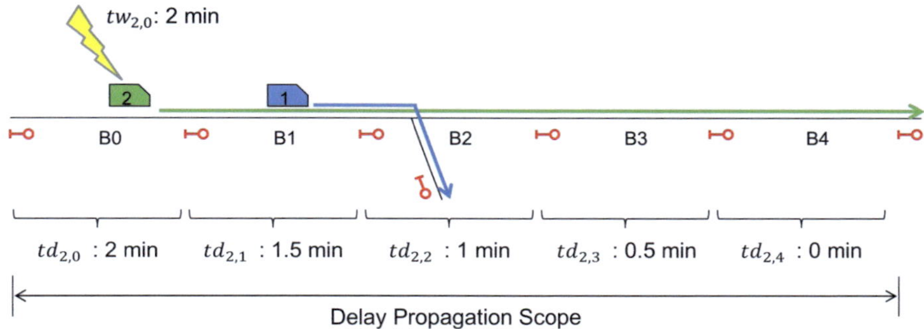


**Figure 4-1: An Example of Delay Propagation Scope (source: [Martin and Liang, 2017])**

The delayed train Z2 may also hinder other trains on its further block section, which would result in knock-on delays of the affected trains. The delay will propagate along the paths of the affected trains and enlarge the delay propagation scope. The basic logic of delay propagation can be summarized as follows: the knock-on delay of a train leads to the delay of the train itself, and the delay of the train itself can lead to knock-on delays of other trains.

From the basic logic of delay propagation, it can be seen that the overall delay propagation scope is the union of the delay propagation scopes of the hindered trains. During the simulation process of the predicted traffic condition, once a train on a certain block section is hindered, the hindered train will then be traced on its further path until one of the following criteria is met:

---

of the macroscale model, many conflicts are ignored (especially in the interlocking areas), so the macroscale model is not recommended for this task.

– The delay caused by the hindrance[15] is eliminated

– The train run or the prediction period is terminated

Before the elimination of delay and the termination of train run and the end of prediction period, it is possible that another train further hinders the already-hindered train. In this case, the delay of the hindered train should be updated, whereupon the hindered train will continue to be traced. The infrastructure nodes that a hindered train passes will be recorded during the tracing process, and output as the delay propagation scope for this hindered train at the end of the simulation. As stated above, the union of the delay propagation scopes of all hindered trains is the overall delay propagation scope.

The dispatching algorithm integrated in the simulation model has a direct influence on delay propagation, because dispatching actions are capable of changing the conflict relationship between trains during the simulation process. However, the description level of each infrastructure node (microscopic, mesoscopic and macroscopic level) should be established primarily according to the significance value, and then the dispatching optimization algorithm can be executed based on the multi-scale simulation model (details of the dispatching optimization algorithm see Chapter 6). In other words, the final dispatching actions are unknown when the delay propagation scope needs to be estimated. In order to solve this contradiction, the dispatching principle FCFS (First Come First Served) is used to assist the estimation of delay propagation scope. This delay propagation can also be regarded as the upper bound of the delay propagation scope.

Based on the delay propagation scope, the infrastructure nodes in the investigated area can be roughly divided into two groups: the nodes inside and the nodes outside the delay propagation scope. The nodes outside of the scope have higher priority to be aggregated than the nodes inside of the scope. In order to obtain more specific

---

[15] The aim of determination of the delay propagation scope is to let the areas relevant to conflicts have a higher priority to be simulated more accurately, so that the influence of a knock-on delay on further conflicts can be more accurately assessed. The influence of a knock-on delay is an important indicator of the dispatching optimization algorithm to be described in Section 6.1.1. So only the delays caused by knock-on delays are considered. Moreover, initial and original delays are controlled variable for dispatching tasks, so they will not be taken into account herein.

ranking of the priorities, the nodes belonging to the same group need to be further sorted according to their aggregation accuracy. In each group, the node with higher aggregation accuracy will receive a higher aggregation priority. Because the knock-on delay is an important output of the simulation model, which is used both by the dispatching optimization module (see Chapter 6), the aggregated accuracy is measured in terms of variation of total knock-on delay in this approach.

The variation of total knock-on delay by aggregating an infrastructure node can be determined with controlled experiments. Generally speaking, the controlled experiment separates search subjects into a controlled group and an experimental group. All variables are kept constant in the controlled group, while a certain investigated variable is changed in the experimental group. By comparing the experimental results of the two groups, the effects of varying the investigated variable can be quantified. To determine the aggregation accuracy, all infrastructure nodes are kept on the microscopic level in the controlled group, and only the investigated infrastructure node is abstracted onto the mesoscopic level. The timetable and stochastic deviations during the operation process are kept the same for both infrastructure scenarios. For each infrastructure scenario, a series of operational simulations should be executed, and the average total knock-on delay can be calculated based on the simulation protocol. The absolute value of the difference between the average total knock-on delays of the two scenarios is the aggregation accuracy of the investigated infrastructure node (Formula (4-1)).

$$ACC_N^{micro \to meso} = \left| \sum_{j=1}^{n_{ges}} \sum_{i=1}^{i=Z_j} tw_{j,i}^{CG} - \sum_{j=1}^{n_{ges}} \sum_{i=1}^{i=Z_j} tw_{j,i}^{EG} \right| \qquad (4\text{-}1)$$

Notation used:

$ACC_N^{micro \to meso}$:   Aggregation accuracy of infrastructure node N from the microscopic level to the most detailed mesoscopic level

$tw_{j,i}^{CG}$, $tw_{j,i}^{EG}$:   Knock-on delay of train $j$ on block section $i$ in the controlled group (CG) or experimental group (EG)

$n_{ges}$:   Total number of trains

$zj$:  Amount of block sections along the path of train $j$

Through the controlled experiment stated above, the aggregation accuracy of abstracting any infrastructure node from the microscopic level to a concrete mesoscopic level can be precisely calculated. However, the mesoscopic level of the multi-scale model is characterized by continuous scaling, which results in a large amount of variants of the mesoscopic infrastructure network, if the investigated area is large. By combining any two adjacent occupation units (see Section 3.2.2.2) a new variant of the mesoscopic infrastructure network will be generated. Obviously, if the aggregation accuracy of the continuous scaling on the mesoscopic level is still evaluated with the controlled experiment, the computational effort will become unfeasible. Furthermore, the loss of accuracy caused by aggregation from the highly detailed mesoscopic level to a more abstracted mesoscopic level is limited compared to that caused by aggregation from microscopic level to the most detailed mesoscopic level. From the microscopic level to the most detailed mesoscopic level (see Section 3.2.2.1), both train movement behavior and regulation of signaling systems are simplified. From the most detailed mesoscopic level to a more abstracted mesoscopic level (see Section 3.2.2.2), the combination of adjacent occupation units mainly changes the releasing times of the involved occupation units. Therefore, it is decided that the controlled experiment only be used to calculate the aggregation accuracy of the infrastructure node abstracted from the microscopic level to the most detailed mesoscopic level, and a simpler and more computationally efficient method was developed to estimate the aggregation accuracy of the continuous scaling on the mesoscopic level, which will be elaborated separately in Section 4.2. As an important result of the method, the aggregation accuracy of combining two adjacent occupation units is returned (Formula (4-2)).

$$ACC_{R_k+R_h}^{meso} = |\Delta E(T^{OVLP,R_k+R_h})| \qquad (4\text{-}2)$$

Notation used:

$ACC_{R_k,R_h}^{meso}$:  Aggregation accuracy of two occupation units $R_k$ and $R_h$ on the mesoscopic level

$\Delta E(T^{OVLP,R_k+R_h})$:      The relative change of expected total overlapping time period caused by combination of $R_k$ and $R_h$ (details see Section 4.2)

With regards to conflicts and the aggregation accuracy, the significance value of each infrastructure node (abstracted from the microscopic level to the most detailed mesoscopic level) and the significant value of a larger mesoscopic occupation unit composed of two other occupation units (abstracted from a certain mesoscopic level to a more abstracted mesoscopic level) can be calculated as follows:

$$SV_N^{micro \to meso} = \chi_{\{l|l \in S_p\}}(N) \cdot M - ACC_N^{micro \to meso} \tag{4-3}$$

$$SV_{R_k+R_h}^{meso} = \chi_{\{l|l \in S_p\}}(N) \cdot M - ACC_{R_k+R_h}^{meso} \tag{4-4}$$

Notation Used:

$SV_N^{micro \to meso}$:      Significance value of an infrastructure node N to be abstracted from the microscopic level to the most detailed mesoscopic level

$SV_{R_k,R_h}^{meso}$:      Significance value of a large mesoscopic occupation unit composed of $R_k$ and $R_h$ on the mesoscopic level

$\chi_{\{l|l \in S_p\}}(N)$:      Indicator function, if the infrastructure node N belongs to the delay propagation scope $S_p$, it is equal to 1, otherwise 0.

$M$:      A number that is at least larger than the maximum of the aggregation accuracies

Based on the significance values, the description levels of infrastructure nodes can be determined in two steps. In the first step, the infrastructure nodes are sorted according to their significance values (Formula (4-3)). The optimal number of occupation units on each description level is limited by computational capacities, time constraints and other on-site factors. This is not within the scope of this dissertation, and it is acceptable to use assumed values for the explanation of the method. Depending on the optimal amount of occupation units on each description level and the ranking of the significance values of infrastructure nodes, the amount of infrastructure nodes kept on the microscopic, the most detailed mesoscopic and macroscopic levels can be determined (Figure 4-2).

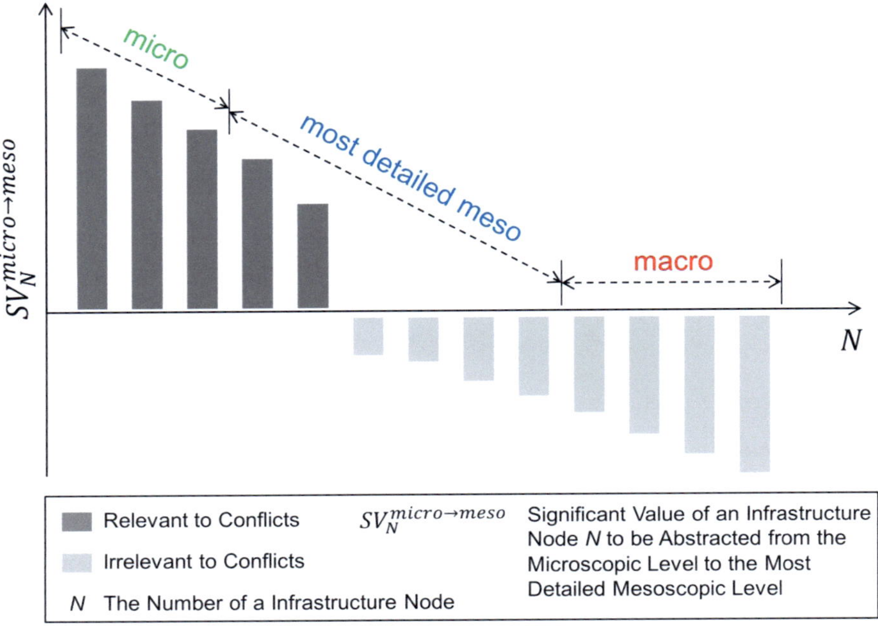



**Figure 4-2: Determination of the Description Level of Each Infrastructure Node (I) (source: [Martin and Liang, 2017])**

In the second step, by further aggregation of occupation units on the mesoscopic level, it may be possible to allow the less significant infrastructure nodes to be simulated on the mesoscopic level instead of on the macroscopic level. The occupation units should be aggregated in sequence according to their significance values calculated with Formula (4-4). When no more infrastructure nodes are able to be transformed from the macroscopic level to the microscopic level, the aggregation process stops. Eventually, the number of infrastructure nodes on each description level as well as the specific form of each infrastructure node (especially on the mesoscopic level) is obtained (Figure 4-3).

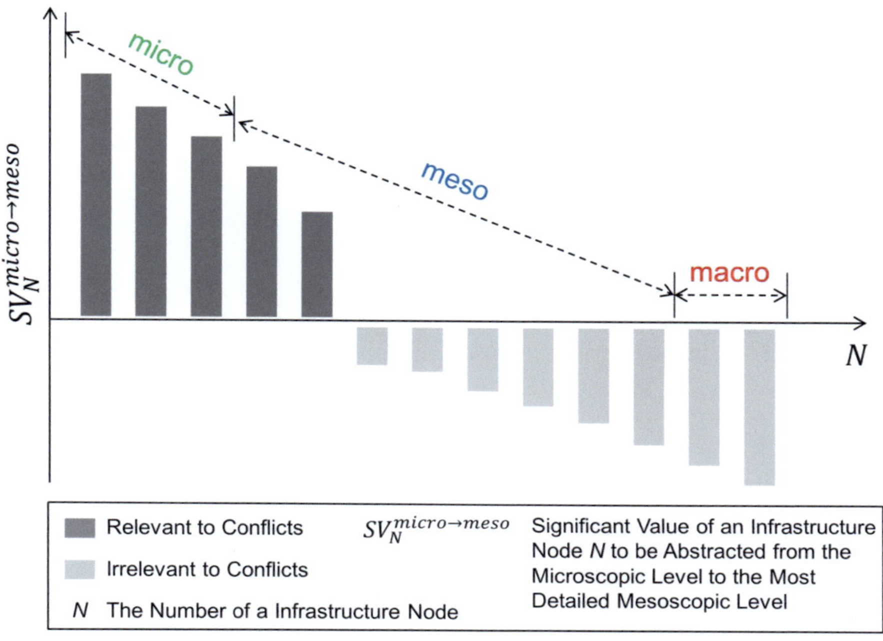


**Figure 4-3: Determination of the Description Level of Each Infrastructure Node (II) (source: [Martin and Liang, 2017])**

## 4.2 Further Aggregation on the Mesoscopic Level

On the mesoscopic level, the basic structures or occupation units can be further aggregated in order to let more infrastructure nodes be simulated on the mesoscopic level. Aggregation of basic structures may influence the conflicts between trains, since the blocking times of train runs on the concerned basic structures is changed. An example is shown in Figure 4-4: two basic structures are to be aggregated, and two train runs are arranged on them. Before aggregation, the conflict is presented as the overlapping of blocking times of different trains. After aggregation, due to the loss of partial releasing, the blocking time of Train Run 1 is correspondingly adjusted. Depending on the throughput of the new occupation unit, the conflict is accordingly enlarged (throughput = 1) or ignored (throughput = 2).

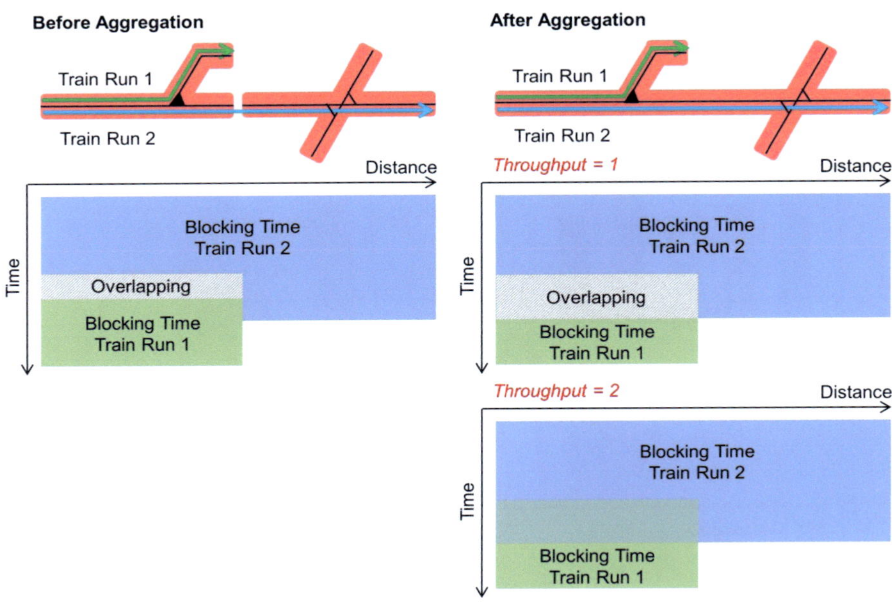


**Figure 4-4: Comparison of Overlapping of Blocking Times Before and After Aggregation (source: [Martin and Liang, 2017])**

It can be seen that the relative change of conflicts – relative change of overlapping of blocking times - between train runs before and after aggregation is a good indicator to quantify the aggregation accuracy. The relative change of a conflict may consequently propagate and influence the other train runs in the same manner as delay propagation. Delay propagation can be accurately estimated with the method described in Section 4.1. However, the amount of aggregation possibilities is large even on a small infrastructure network, and the computational effort becomes unfeasible if the propagation is also taken into account. For practical applications, the relative change of conflict will be minimized from the source, and its further influences will be ignored.

The blocking time of a train run on a certain occupation unit is described with two variables: start blocking time and length of the blocking time. In case of hindrances (e.g. unscheduled stop before a red signal) the blocking time may be extended, and the average extension can be determined through a large amount of operational simulations. The summation of the length of a scheduled blocking time and its average extension is the expected length of the actual blocking time, which is adopted to estimate overlapping times. In the following context, if not otherwise stated, the length of a blocking time refers to the expected length of the actual blocking time. Because of deviations in real operation, the actual start blocking time can be expressed as:

$$t_{j,i}^{start,Ist} = t_{j,i}^{start,Soll} + td_{j,i} \qquad (4\text{-}5)$$

Notation used:

$t_{j,i}^{start,Ist}$:  Actual start blocking time of train $j$ on occupation unit $i$

$t_{j,i}^{start,Soll}$:  Scheduled start blocking time of train $j$ on occupation unit $i$

$td_{j,i}$:  Delay of train $j$ on occupation unit $i$

In general it can be assumed that $td_{j,i}$ follows Erlang-K distribution. As a special case of Erlang-K distribution, negative exponential distribution was used. Thus, $t_{j,i}^{start,Ist}$ follows the same probability distribution, because $t_{j,i}^{start,Soll}$ is a constant value fixed by schedule.

For the sake of convenience in computerized processing, discrete time is used instead of continuous time. Accordingly, the time-related variables including the start and end time of investigated time period, length of blocking time and start blocking time should be adjusted to become an integral multiple of the discrete interval. Moreover, the probability distribution of delay can be expressed with a frequency histogram using the same time interval (e.g. the left subgraph in Figure 4-5).

Depending on the various possible delays, the actual blocking time occurs at different time periods with certain probabilities. The right subgraph of Figure 4-5 shows an example: for a certain delay x ($x \in \{0, \Delta t, 2\Delta t, 3\Delta t\}$) the start blocking time ($x + t_s$) can be accordingly determined, and the time intervals within the length of blocking time from ($x + t_s$) to ($x + t_s + 3\Delta t$) will be blocked. The probability that blocking time occurs at this time period is equal to the probability of the occurrence of the delay x. Because time is discretized, the probability of the occurrence of the delay x is defined as the probability that the delay belongs to the time interval [x, x+$\Delta t$).

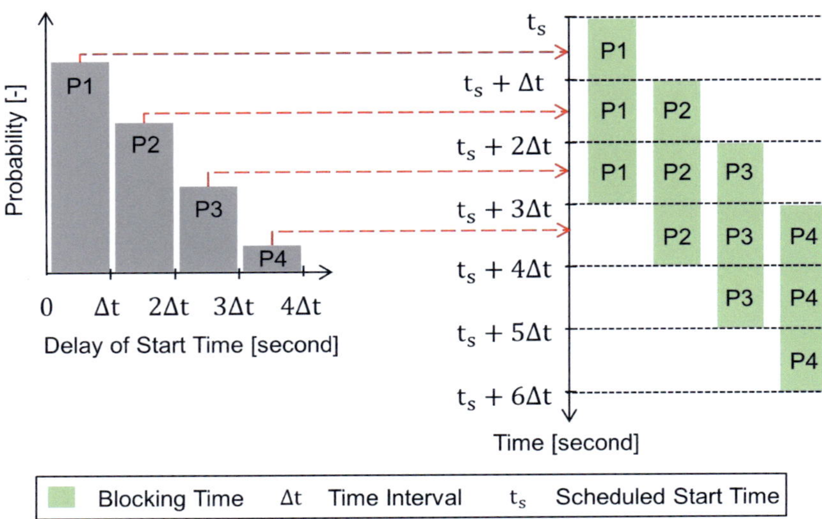


**Figure 4-5: Probability Distribution of Delay and Blocking Time (source: [Martin and Liang, 2017])**

The probability distribution of blocking time can be transformed into the blocking probability of each time interval as shown in Figure 4-6, and the blocking probabilities of the time intervals outside of the range $[t_s, t_s + 6\Delta t)$ should be set to zero in this example.

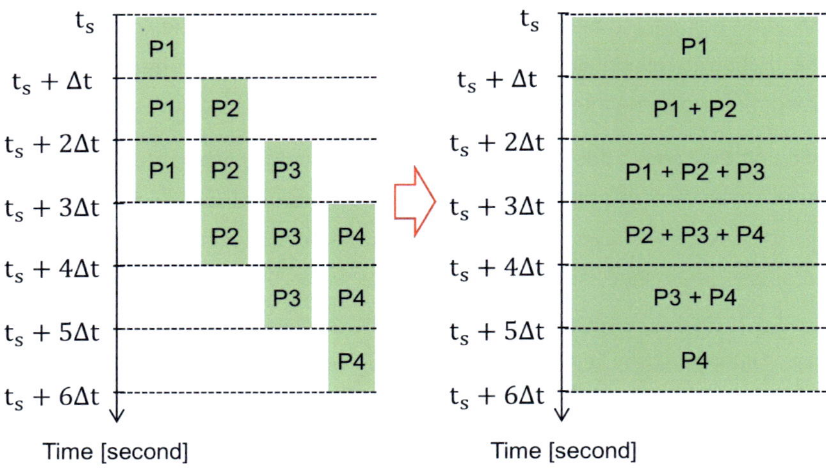


**Figure 4-6: Blocking Probability of Each Time Interval (source: [Martin and Liang, 2017])**

Using the method above, the blocking probability of each time interval can be determined for each involved train run on an occupation unit. Thereby, the expected overlapping of blocking times between train runs in each time interval can be calculated, and the sum of these results is the expected total overlapping in the entire investigated time period, with which the relative changes of conflicts before and after aggregation can be simply calculated. Evidently, the key point is to calculate the expected overlapping in each time interval. To explain the calculation method clearly, a simple example is created in Figure 4-7: there are three train runs Z1, Z2 and Z3 on an occupation unit named R3, and the investigated time period is. from $t_0$ to $t_m$. The blocking probability of each time interval for each train run is given in the second subgraph, and $[t_2, t_3)$ is chosen as a sample time interval.

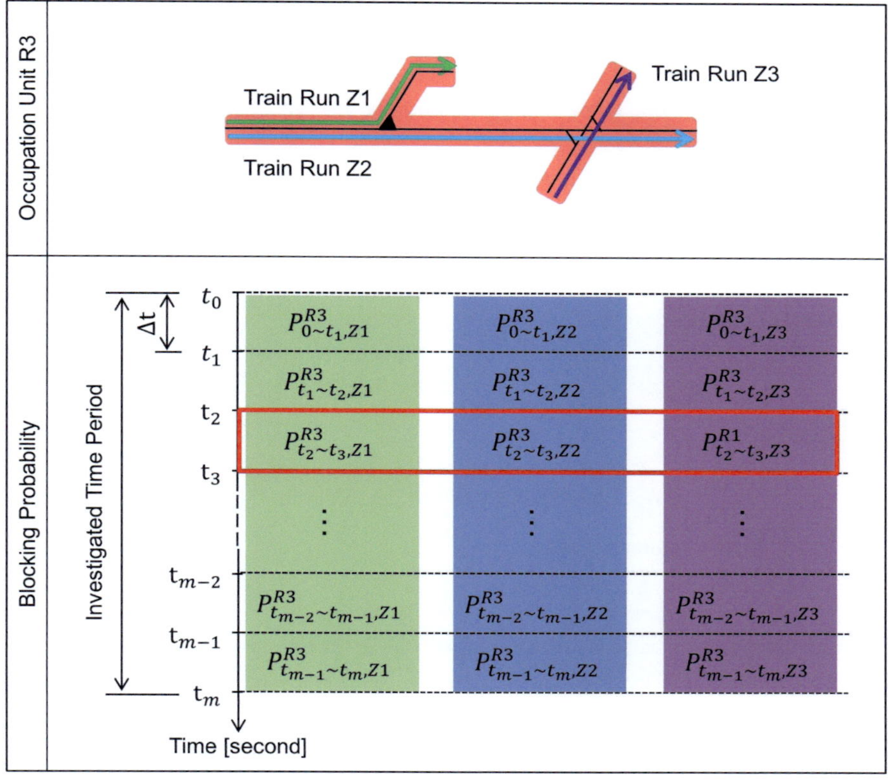


**Figure 4-7: Calculation of Expected Overlapping Time Period on an Occupation Unit (source: [Martin and Liang, 2017])**

On this occupation unit, the conflicts can occur between every two trains or between these three trains. So, there are $(C_3^2 + C_3^3)$ possible conflict situations[16]. Furthermore, because R3 is composed of two basic structures (see Figure 4-4), the throughput[17] of R3 is either 1 or 2.

Take Z1 and Z2 as an example of the conflict situation occurred only between two trains, and the expected overlapping time period between them in the time interval $[t_2, t_3)$ can be calculated with the general Formula (4-6).

$$E(T_{t_2 \sim t_3, Z1, Z2}^{OVLP, R3}) = P_{t_2 \sim t_3, Z1, Z2}^{R3} \cdot t_{t_2 \sim t_3, Z1, Z2}^{OVLP, R3} \tag{4-6}$$

Notation used:

$P_{t_2 \sim t_3, Z1, Z2}^{R3}$:      Occurrence probability of the conflict situation between Z1 and Z2 on the occupation unit R3 in the time interval from $t_2$ to $t_3$

$t_{t_2 \sim t_3, Z1, Z2}^{OVLP, R3}$:      Total overlapping time period between Z1 and Z2

$E(T_{t_2 \sim t_3, Z1, Z2}^{OVLP, R3})$:      Expected total overlapping time period between Z1 and Z2

The occurrence probability of the corresponding conflict situation can be easily calculated based on the blocking probability of each time interval (e.g. second subgraph in Figure 4-7) as follows:

$$P_{t_2 \sim t_3, Z1, Z2}^{R3} = P_{t_2 \sim t_3, Z1}^{R3} \cdot P_{t_2 \sim t_3, Z2}^{R3} \cdot (1 - P_{t_2 \sim t_3, Z3}^{R3}) \tag{4-7}$$

Where:

$P_{t_2 \sim t_3, Zj}^{R3}$:      Blocking probability of the time interval $[t_2, t_3)$ for $Zj$ on the occupation unit R3

---

[16] If $n$ train runs exist on a certain occupation unit, the amount of possible conflict situations are $(C_n^2 + C_n^3 + \cdots + C_n^n)$. To calculate the overlapping time period in a certain time interval, all possible conflict situations should be considered.

[17] In general, a larger occupation unit (designated as R3) is composed of another two occupation units (designated as R1 and R2). The throughputs of R1 and R2 are $TP_{R1}$ and $TP_{R2}$ respectively. The range of $TP_{R3}$ is [1, $TP_{R1}+TP_{R2}$ ]

The total overlapping time period in the corresponding conflict situation changes as the throughput varies. In case the throughput is 1, the total overlapping time period between Z1 and Z2 is

$$t^{OVLP,R3}_{t_2 \sim t_3, Z1, Z2} \big|_{TP_{R3}=1} = \Delta t \tag{4-8}$$

In case the throughput is 2, the overlapping between these two trains will be ignored.

$$t^{OVLP,R3}_{t_2 \sim t_3, Z1, Z2} \big|_{TP_{R3}=2} = 0 \tag{4-9}$$

The expected overlapping time between Z1 and Z2 can be calculated:

$$E(T^{OVLP,R3}_{t_2 \sim t_3, Z1, Z2}) \big|_{TP_{R3}=1} = P^{R3}_{t_2 \sim t_3, Z1} \cdot P^{R3}_{t_2 \sim t_3, Z2} \cdot (1 - P^{R3}_{t_2 \sim t_3, Z3}) \cdot \Delta t \tag{4-10}$$

$$E(T^{OVLP,R3}_{t_2 \sim t_3, Z1, Z2}) \big|_{TP_{R3}=2} = P^{R3}_{t_2 \sim t_3, Z1} \cdot P^{R3}_{t_2 \sim t_3, Z2} \cdot (1 - P^{R3}_{t_2 \sim t_3, Z3}) \cdot 0 \tag{4-11}$$

On the condition that conflicts occur between these three trains, the occurrence probability of this conflict situation is

$$P^{R3}_{t_2 \sim t_3, Z1, Z2, Z3} = P^{R3}_{t_2 \sim t_3, Z1} \cdot P^{R3}_{t_2 \sim t_3, Z2} \cdot P^{R3}_{t_2 \sim t_3, Z3} \tag{4-12}$$

In case the throughput is 1, the blocking times of every two of these three trains overlapped (there are $C_3^2$ possibilities). So the total overlapping time period is

$$t^{OVLP,R3}_{t_2 \sim t_3, Z1, Z2, Z3} \big|_{TP_{R3}=1} = C_3^2 \cdot \Delta t = 3 \cdot \Delta t \tag{4-13}$$

The throughput indicates the number of trains that can block an occupation unit without conflict. So the overlapping time periods occurring between the trains belonging to the throughput can be ignored[18]. When the throughput is 2, the total overlapping time period should be adjusted as follows:

$$t^{OVLP,R3}_{t_2 \sim t_3, Z1, Z2, Z3} \big|_{TP_{R3}=2} = (C_3^2 - C_2^2) \cdot \Delta t = 2 \cdot \Delta t \tag{4-14}$$

The expected overlapping times among Z1, Z2 and Z3 are:

---

[18] Generally speaking, if $n_{cfl}$ trains are involved in a certain conflict situation and the throughput is $n_{TP}$, the number of overlapping is max $\{0, (C^2_{n_{cfl}} - C^2_{n_{TP}})\}$. $n_{cfl}$ is always equal or larger than 2, so $C^2_{n_{cfl}}$ is equal or larger than 1. However, $n_{TP}$ can equal to 1. In this case $C^2_1$ is equal to 0 by definition.

$$E\left(T_{t_2\sim t_3,Z1,Z2,Z3}^{OVLP,R3}\right)\big|_{TP_{R3}=1} = P_{t_2\sim t_3,Z1}^{R3} \cdot P_{t_2\sim t_3,Z2}^{R3} \cdot P_{t_2\sim t_3,Z3}^{R3} \cdot 3 \cdot \Delta t \qquad (4\text{-}15)$$

$$E\left(T_{t_2\sim t_3,Z1,Z2,Z3}^{OVLP,R3}\right)\big|_{TP_{R3}=2} = P_{t_2\sim t_3,Z1}^{R3} \cdot P_{t_2\sim t_3,Z2}^{R3} \cdot P_{t_2\sim t_3,Z3}^{R3} \cdot 2 \cdot \Delta t \qquad (4\text{-}16)$$

The expected total overlapping time period in the time interval $[t_2, t_3)$ is the sum of the results of all possible conflict situations.

$$
\begin{aligned}
E\left(T_{t_2\sim t_3}^{OVLP,R3}\right)\big|_{TP_{R3}=1\ or\ 2} ={}& E\left(T_{t_2\sim t_3,Z1,Z2}^{OVLP,R3}\right)\big|_{TP_{R3}=1\ or\ 2} \\[2ex]
&+ E\left(T_{t_2\sim t_3,Z2,Z3}^{OVLP,R3}\right)\big|_{TP_{R3}=1\ or\ 2} \\[2ex]
&+ E\left(T_{t_2\sim t_3,Z1,Z3}^{OVLP,R3}\right)\big|_{TP_{R3}=1\ or\ 2} \\[2ex]
&+ E\left(T_{t_2\sim t_3,Z1,Z2,Z3}^{OVLP,R3}\right)\big|_{TP_{R3}=1\ or\ 2}
\end{aligned}
\qquad (4\text{-}17)
$$

The expected total overlapping time period in the other time intervals can be calculated with the same process as elaborated above. Eventually, the expected total overlapping time period in the whole investigated time period can be calculated, which is the sum of the results of all time intervals.

$$
\begin{aligned}
E\left(T^{OVLP,R3}\right)\big|_{TP_{R3}=1\ or\ 2} ={}& E\left(T_{t_0\sim t_1}^{OVLP,R3}\right)\big|_{TP_{R3}=1\ or\ 2} \\[2ex]
&+ E\left(T_{t_1\sim t_2}^{OVLP,R3}\right)\big|_{TP_{R3}=1\ or\ 2} \\[2ex]
&\vdots \\[2ex]
&+ E\left(T_{t_{m-1}\sim t_m}^{OVLP,R3}\right)\big|_{TP_{R3}=1\ or\ 2}
\end{aligned}
\qquad (4\text{-}18)
$$

To estimate the aggregation accuracy for the two adjacent occupation units R1 and R2, the expected total overlapping time periods on them before aggregation should be determined with the same process in advance. The only difference is that the throughputs of R1 and R2 are pre-given. The relative change of the expected total overlapping time period can be calculated as follows.

$$\Delta E\left(T^{OVLP,R3}\right)\big|_{TP_{R3}=1\ or\ 2} = \ E\left(T^{OVLP,R1}\right)\big|_{TP_{R3}=1\ or\ 2}$$

$$+E\left(T^{OVLP,R2}\right)\big|_{TP_{R3}=1\ or\ 2} \qquad (4\text{-}19)$$

$$-E\left(T^{OVLP,R3}\right)\big|_{TP_{R3}=1\ or\ 2}$$

By comparing the absolute values of the relative changes in case of the different throughputs, the lower absolute value will be chosen to indicate the aggregation accuracy of R3, and the corresponding throughput will be used as the representative throughput of R3.

The calculation procedure explained above can easily be generalized to estimate the aggregation accuracy of any larger occupation unit. The procedure can be summarized into four steps in sequence:

1) Calculate the expected overlapping time periods on two to-be-aggregated occupation units;

2) Adjust the blocking times of the concerned trains on the larger aggregated occupation unit and recalculate the blocking probabilities of each time interval for the trains;

3) Calculate the expected overlapping time period on the larger aggregated occupation unit;

4) Calculate the relative change of overlapping time period before and after aggregation, and determine the aggregation accuracy and representative throughput for the larger occupation unit.

In the further up-scaling process, the occupation unit with higher aggregation accuracy (a lower relative change of expected overlapping time period) will receive a higher aggregation priority.

# 5  Multi-scale Simulation Model with Priority Sequence Control

The multi-scale simulation model depicted in Chapter 3 adopts implicitly the First Come First Serve (FCFS) principle. The resource requirements of trains are attended in the order that they arrive, without other sequence control constraints. The train priority sequences on relevant infrastructure resources are unknown before the end of simulations. The Banker's algorithm is implemented for the purpose of deadlock avoidance. The simulation model is capable of generating deadlock-free timetables independently, which are used as the basic timetable in further dispatching optimization processes.

The basic timetable generated by the simulation model without train priority sequence control will be modified by the dispatching optimization module through a series of dispatching methods (see Chapter 6); as a result a dispatched timetable is generated, which is represented as train priority sequence lists on relevant infrastructure resources. In order to simulate dispatched timetables and provide necessary data for timetable characteristics evaluation in the optimization process, a simulation model with train priority sequence control was designed as part of the DFG project [Martin and Liang, 2017]. In this model, train runs are regulated explicitly by a pre-given dispatched timetable. Because Banker's algorithm could change train paths and train priority sequences on infrastructure resources, it is not applied in the simulation model with priority sequence control, in order to avoid the control contradictions. The interconnection between the simulation model and the dispatching optimization model is illustrated in Figure 5-1. The multi-scale simulation model with train priority sequence control developed in [Martin and Liang, 2017] will be described in this chapter, and the dispatching optimization algorithm will be described in Chapter 6.

Essentially, the train priority sequence control and the dispatching optimization algorithm on the macroscopic level follow the same principle as those on the microscopic and mesoscopic levels. Nevertheless, due to the high level of abstraction of the macroscopic level, the infrastructure node – loop non track do not exist anymore on this level (for more details of the infrastructure hierarchy on the microscopic and mesoscopic level it is referred to Figure 3-30). Accordingly, the algorithms of train priority sequence control and dispatching optimization need to be fine-tuned on the macroscopic level. For instance, it is possible that knock-on delays occur on the area of loop non track on the microscopic and mesoscopic levels, but not possible on the macroscopic level. Therefore, the algorithm details associated with loop non track should be removed on the macroscopic level. Because the same principle is applied, the algorithms of priority sequence control and dispatching optimization on the macroscopic level can be easily derived from those on the microscopic and mesoscopic levels. To avoid content duplication, the algorithms of priority sequence control and dispatching optimization will be only discussed on the microscopic and mesoscopic levels in Chapter 5 and Chapter 6.

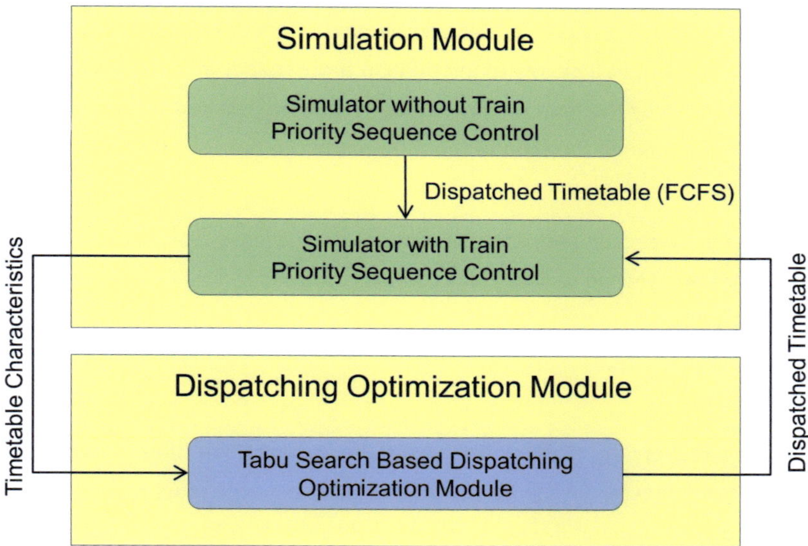


**Figure 5-1: Interconnection between Simulation Module and Dispatching Optimization Module**

## 5.1 Basics of Priority Sequence Control

In principle, the workflow of the simulation model with priority sequence control is the same as the workflow of the simulation model without priority sequence control. Likewise the workflow includes three procedures in a single processing step at a time interval: request resources, allocate resources and proceed with simulation tasks. In each procedure, a new mechanism of priority sequence control is added. The infrastructure hierarchy defined in Section 3.2.1 is used herein, and train priority sequence constraints are added onto the infrastructure nodes to meet dispatching-specific requirements. Train priority sequence control is realized through an arrival sequence list and a departure sequence list on each loop track and open track section (except the open track section, only including virtual block sections). The two lists are initially identical. As a train occupies or releases a loop track or an open track section, the arrival list or departure list on it is updated. The details of the control mechanism in the three procedures will be separately elaborated in Section 5.2 and 5.3.

## 5.2    Priority Sequence Control in the Resource Requirement Procedure

In the procedure of resource requirements, new resource requirements should be checked according to the priority sequence control constraints. To check a new requirement of a train, two pieces of information are needed: the last physically occupied block section by the train (named Block 1) and the required block section which is beyond Block 1 along the path of the train (named Block 2). The node attribute of Block 1 can be any of the following types:

- Open track section (virtual block sections not included)
- Open track section (virtual block sections included)
- Loop track
- Loop non track
- Junction

The node corresponding to Block 1 is called the current node. If the current node is any of the last two types in the list above, the new resource requirement is certainly safe and the permission of the new resource requirement will be given directly. When a train physically arrives at a block section located in the area of loop non track or junction, it has already fulfilled the sequence control constraints on the subsequent open track section or loop track in advance. Otherwise, the train should stop on an open track section or loop track and wait until the entry permission of the subsequent open track section or loop track is given. An example of this situation is shown in Figure 5-2. Train 1 can enter the area of the loop non track, because the subsequent loop track has already been reserved for it. On the contrary, Train 2 is required to wait on the open track section, because the subsequent loop track is reserved for it in a stacked position. The case that the current node is a junction follows the same principle.

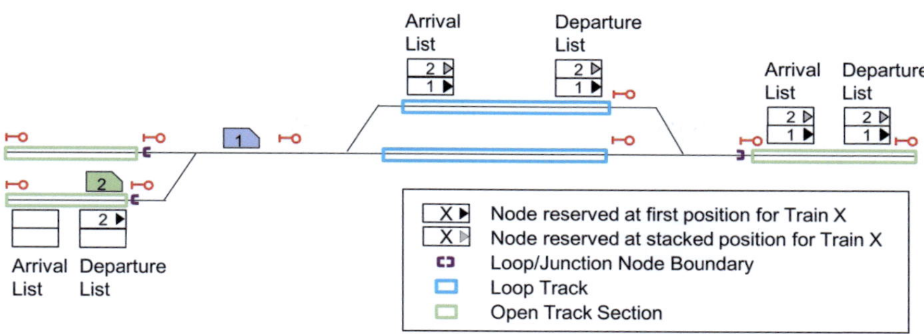


**Figure 5-2: An Example of Unnecessary Priority Sequence Control**

If the node attribute of Block 1 is any of the first three types, the new requirement may violate existing constraints and should be proven, barring two exceptions. The first exception depends on the position of Block 2. Based on the included basic structures of Block 2, whether Block 2 is completely included in the current node can be determined. If Block 2 is completely included in the current node, the train will not leave the current node in the current time interval, which means that the new resource requirement has no influence on the priority sequence constraints, and can be permitted directly. The second exception depends on the type of the subsequent loop track or open track section. If the subsequent section is an open track containing only a virtual block section, the new resource requirement can be permitted directly. Open track sections containing only virtual block sections do not have the function of priority sequence control. An example of these two exceptions is shown in Figure 5-3. Train 1 will enter the virtual block section, and the new resource requirement will be permitted directly. Train 2 requires the next block section, but will not leave the current open track section in the current time interval. Therefore, the new resource requirement will also be permitted directly.

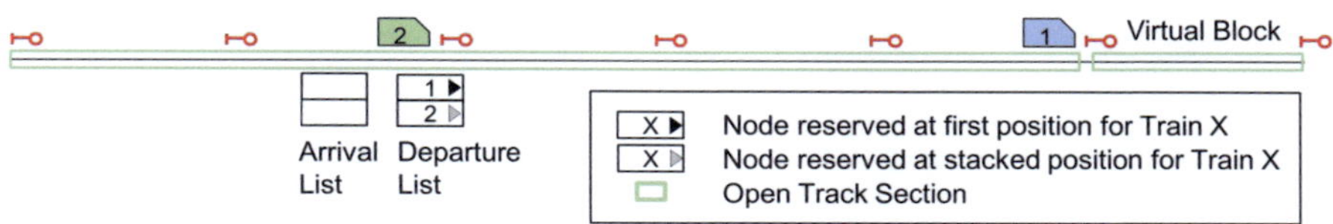


**Figure 5-3: An Example of Two Exceptions in the Priority Sequence Control (source: [Martin and Liang, 2017])**

Except for the situations described above, the new resource requirement must be proven with the following steps:

Step 1: Determine the subsequent node (i.e. loop track or open track section) based on the current node.

Step 2: If the subsequent node is not reserved at the first position for the considered train in the arrival list, go directly to Step 6. Otherwise continue to the next step.

Step 3: If the subsequent node is reserved at the first position for this train in the departure list, go to Step 5. Otherwise continue to the next step.

Step 4: If the running direction of the train that departs previously to this train in the departure list of the subsequent node is opposite to the running direction of this train, go to Step 6. Otherwise continue to Step 5.

Step 5: Permit the new resource requirement.

Step 6: Reject the new resource requirement

The complete flow of train priority sequence control in the procedure of resource requirement is shown in Figure 5-4.

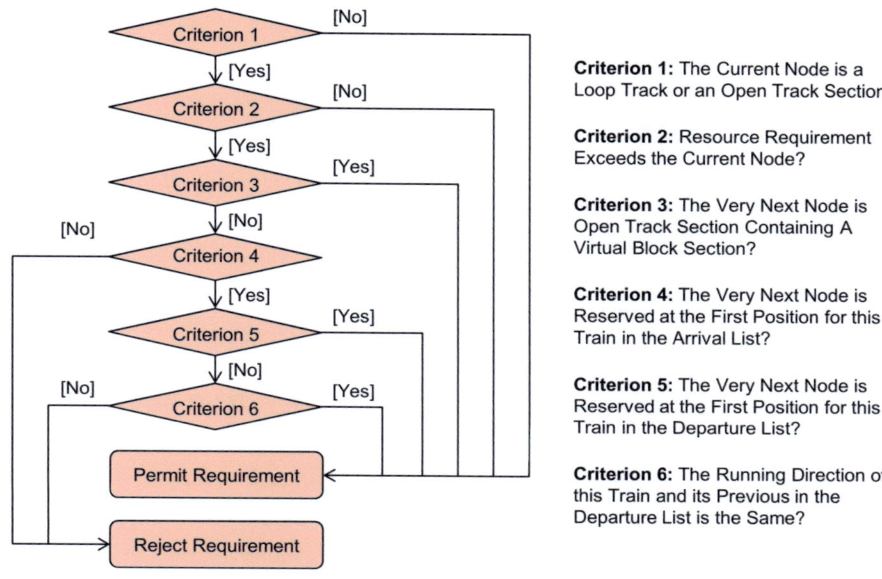


**Figure 5-4: The Complete Flow of Priority Sequence Control in Resource Requirement Procedure**

## 5.3 Priority Sequence Control in the Resource Allocation and Proceeding with Simulation Tasks

In the procedure of resource allocation, the new resource requirement should pass the conflict-free test and the deadlock-free test in the simulation model without priority sequence control. However, the pre-given train priority sequence lists on each loop track and open track section in the simulation model with priority sequence control have already passed the deadlock-free test before simulation starts as described in Section 6.1.3. Therefore, the new resource requirement of a train only needs to be conflict-free, which means at the time point of occurrence of a new resource requirement, the required resources must not be occupied by the other trains. If a new resource requirement has passed the conflict-free test, the required resources will be directly allocated to the corresponding train.

In the procedure of proceeding with simulation tasks, besides the simulation tasks described in Section 3.1.2.3, updating of priority sequence constraints (updating of arrival and departure lists on loop tracks and open track sections) is additionally implemented. This task will be executed after the position of a train is updated in each time interval. The updating of arrival lists depends on the new resource requirement, and the updating of departure lists depends on the released resources in one time interval. Thus, this task requires two pieces of information as inputs, and is correspondingly divided into two subtasks:

- Updating of arrival lists

    Input: new resource requirements without consideration of overlap

- Updating of departure lists

    Input: released resources in one time interval

Only requested resources other than overlap are considered for the updating of arrival lists. In most cases an overlap is occupied together with its corresponding block sections. However, in case of scheduled or unscheduled stops before a main signal of a block section, the corresponding overlap will be released after the train stops. The repeated changes of overlap occupation are not suitable for train priority sequence control, so only the new allocated resources, without overlaps, is considered. Furthermore, the new resource requirements used here are only the requirements generated by the resource requirement procedure that passed the conflict-free test and was successfully allocated, since only changes of resource occupation may influence arrival lists.

For updating of arrival lists, the node attributes of the basic structures included in a new resource requirement without overlaps are collected. If any of the nodes is a loop track or an open track section, and the first position of its arrival list is reserved for the requester, the train will be removed from the arrival list. For updating of departure lists, the node attributes of the released basic structures will be collected. If any of nodes is a loop track or open track section, the train will be removed from its departure list, except when the train is still occupying the node.

# 6 Domain-specific Dispatching Optimization Model

The dispatching optimization model developed in this dissertation is used to adjust the train priority sequences on loop tracks and open track sections through a series of dispatching actions in order to find an optimal dispatched timetable. The dispatching optimization in railway operation is a typical combinatorial optimization, and for such problems exhaustive search is not feasible on large scale cases. A metaheuristic algorithm is capable to balance the quality of solution and the computational complexity. A widely used metaheuristic algorithm - tabu search - is preferred as the basis of the dispatching optimization model. Tabu search (TS) algorithm is a local search based metaheuristic algorithm firstly proposed by Glover in 1986 [Glover, 1986]. Tabu search follows the neighborhood search procedure as local search, iteratively moving from an initial solution to a new neighbor solution in its neighborhood.

The framework of the tabu search algorithm implemented in this approach is shown in Figure 6-1. The optimization process starts with an initial solution, which can be generated by the simulation model developed in Chapter 3. By solving of an inherent conflict in the initial solution, a new neighbor solution is accordingly generated. Because simulation of candidate solutions is a time-consuming task, only a subset of the neighborhood has the chance to be added in the candidate list. In this approach, both intensification and diversification elements are included in the candidate list, which will be elaborated in Section 6.1. After candidate list was constructed, the candidate will be simulated and evaluated with the pre-defined dispatching objective function (see Section 6.2). Under the constraint of tabu list (see Section 6.3), the best solution will be selected out and used to replace the initial solution. This procedure is performed iteratively until the terminate specification is fulfilled (see Section 6.3). The best solution among all historical results is chosen as the final solution. At the end of this chapter (Section 6.4), experiments on a reference example are conducted.

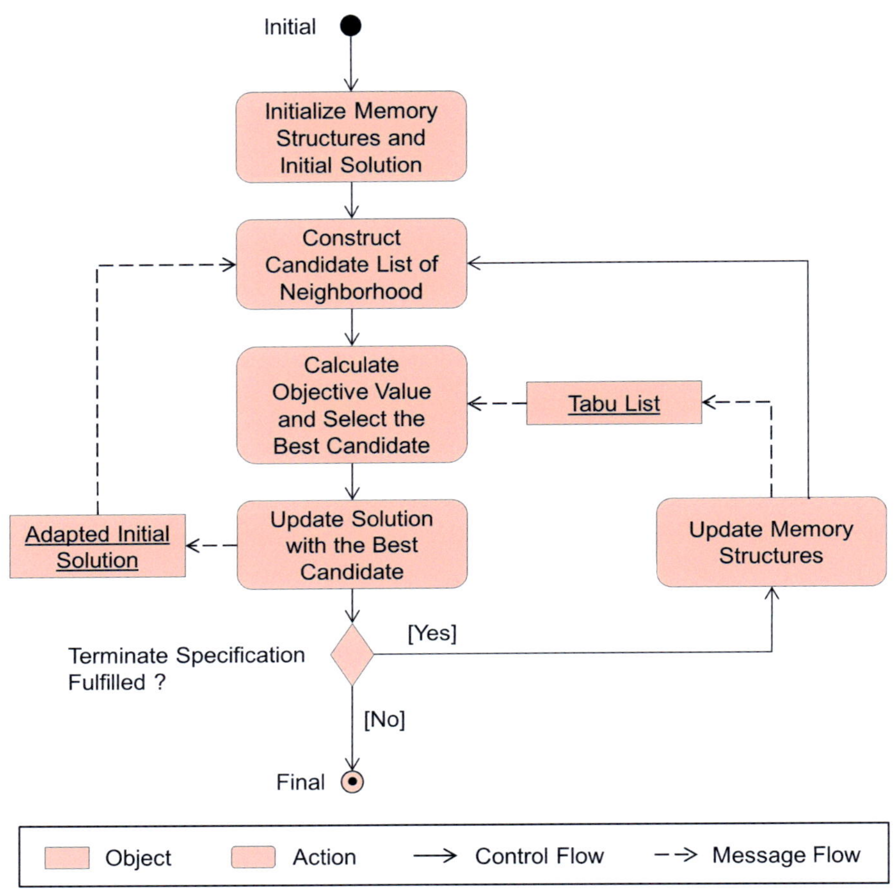


**Figure 6-1: The Framework of the Implemented Tabu Search Algorithm**

## 6.1    Construction of the Candidate List

To construct the candidate list, the knock-on delay in the initial solution will be ranked according to their priority primarily, and by resolving a selected knock-on delay by a series of dispatching actions such as passing, overtaking and replatforming a candidate solution will be generated. Within the scope of the DFG project [Martin and Liang, 2017] a greedy algorithm based dispatching optimization algorithm was developed. For the construction of the candidate set for the greedy algorithm, knock-on delays were also ranked to enable the important conflicts to be resolved primarily, and a comprehensive approach of train priority sequence adjustment (resolution of knock-on delays) were developed in the project. Even though a different optimization algorithm (i.e. tabu search) was employed in this dissertation, the algorithm for calculating the priority of knock-on delays and the approach of train priority sequence adjustment apply equally here. They will be elaborated in Section 6.1.1 and 6.1.2 respectively. The higher ranked knock-on delays are used to generate intensification elements and the left lower ranked knock-on delays are used to generate diversification elements for the candidate list of the tabu search algorithm. The candidate list strategy will be described in detail in Section 6.1.4. Only feasible solutions are considered in the construction of the candidate list, so the solutions that cannot pass the deadlock-free test in Section 6.1.3 will be discarded directly.

### 6.1.1   Priority Calculation for Knock-on Delays

The algorithm for calculating the priority of knock-on delays consists of two criteria: weighted knock-on delay $\widehat{tw}$ and the influence of knock-on delay on further conflicts $Inf_{tw}$. Both of the criteria are dispatching objective function oriented. Weighted knock-on delay reflects the direct contribution of a knock-on delay to the value of the objective function, and the influence of knock-on delay on further conflicts reflects the indirect contribution of a knock-on delay to the value of the objective function.

In the calculation of priority for knock-on delays, knock-on delays are weighted in the same manner as they are weighted in the dispatching objective function[19]. The criteria weighted knock-on delay is calculated with Formula (6-1).

$$\widehat{tw}_{j,i} = \frac{C_j + Zfl}{1 + Zfl} \cdot tw_{j,i} \qquad (6\text{-}1)$$

---

[19] As can be seen in Formula (6-1), the weight of knock-on delay is determined by two parameters – the constant for weighting knock-on delays $C_j$ and the viscosity of dispatching conditions $Zfl$. These two parameters are also included in the dispatching objective function implemented in this approach. Therefore, the calculation methods of these two parameters will be explained in detail in Section 6.2 along with the dispatching objective function (see Formula (6-10) and Formula (6-11)).

Notation used:

$\widehat{tw}_{j,i}$  Weighted knock-on delay of train j on block section i

$C_j$  Constant for weighting knock-on delays of train j

$Zfl$  Viscosity of dispatching conditions

$tw_{j,i}$  Knock-on delay of train j on block section i

Once a train experiences a knock-on delay on a certain block section, the delay will accompany the train on the further block sections along its path until the delay is eliminated either by recovery times or by the termination of the train run. As shown in Figure 6-2, a train Z2 on the block section B0 was hindered by another train Z1 on the block section B1, due to which Z2 obtained a knock-on delay $tw_{2,0}$ of two minutes on B0. The knock-on delay resulted in delays of Z2 $td_{2,i}$ on the further block sections along its path. Because $tw_{2,0}$ is the only source of $td_{2,i}$, $tw_{2,0}$ is entirely responsible for $td_{2,i}$. In order to express more clearly, $tw_{2,0}$ is defined as the source knock-on delay of a certain delay $td_{2,i}$.

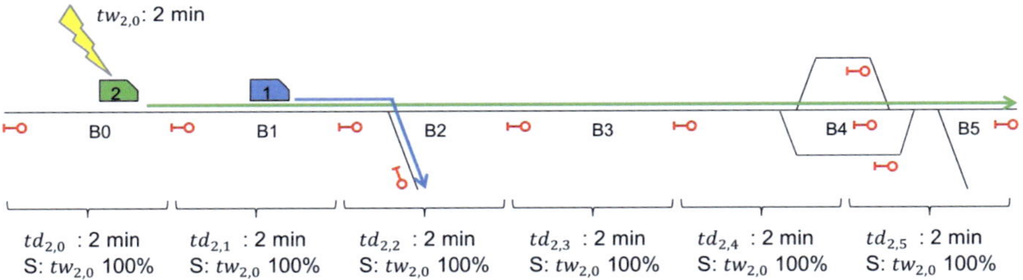


**Figure 6-2: Relationship between Delay and Knock-on Delay of a Train (I) (source: [Martin and Liang, 2017])**

In case a delay is caused by more than one knock-on delay, the responsibility of each knock-on delay will be calculated. As shown in Figure 6-3, Z2 on block section B3 was hindered by another train, Z3, on block section B4, due to which Z2 experienced a knock-on delay $tw_{2,3}$ of three minutes on B3. The knock-on delay $tw_{2,3}$ resulted in the delay of Z2 on B3, B4 and B5 being increased to 5 minutes. In this situation, the delay $td_{2,3}$ or $td_{2,4}$ or $td_{2,5}$ was caused by two sources $tw_{2,0}$ and $tw_{2,3}$, successively. The responsibility for delay $td_{2,3}$ or $td_{2,4}$ or $td_{2,5}$ will, therefore, be distributed proportionally to $tw_{2,0}$ and $tw_{2,3}$. Delay $tw_{2,0}$ assumes 40% responsibility, and delay $tw_{2,3}$ assumes 60% responsibility.

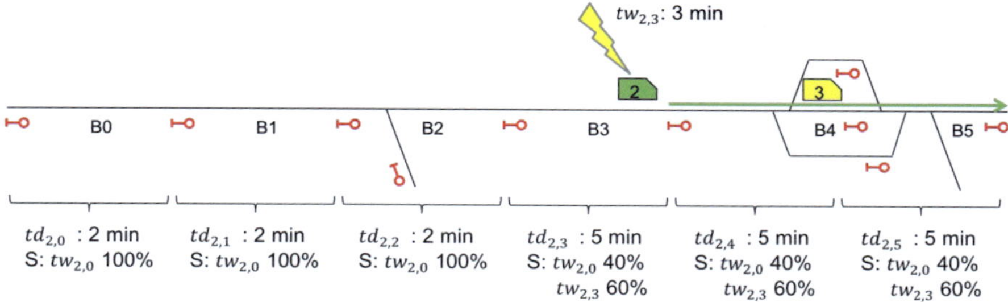


**Figure 6-3: Relationship between Delay and Knock-on Delay of a Train (II) (source: [Martin and Liang, 2017])**

The procedure to describe the relationship between delays[20] and knock-on delays of a train quantitatively is summarized into the following steps:

Step 1: Calculate the knock-on delays of a train. It is supposed that there are $n$ knock-on delays for the considered train. ($n \geq 1$)

Step 2: Select the $i^{th}$ knock-on delay; enumerate the block sections from the one on which the knock-on delay occurred to the second to last block section along the path of the train. ($i \in [1, n]$)

Step 3: For each enumerated block section, a delay instance will be generated if one does not previously exist. A delay instance is identified by two attributes: the delayed train and the block section.

Step 4: Record the $i^{th}$ knock-on delay in the source knock-on delay list of each enumerated block section.

Step 5: If all knock-on delays of the train have been analyzed, proceed to Step 6, otherwise return to Step 2.

---

[20] The delay discussed here includes only delays caused by knock-on delays in the investigation area. The original delay of a train at the initial station in the investigation area and the initial delay of a train at the boundary of the investigation area are not considered, since original delays and initial delays are controlled variable for dispatching tasks.

Step 6: Analyze each block section along the path of the train. If a delay instance is defined on a block section, the responsibility of each source knock-on delay will be calculated with Formula (6-2).

$$P^{Rb}_{td_{j,k}, tw_{j,i}} = \frac{tw_{j,i}}{\sum_{b=1}^{b=n_j^{block}} tw_{j,b} \cdot \chi_{\{l|\, tw_{j,l} \in S^{source}_{td_{j,k}}\}}} \tag{6-2}$$

Notation used:

$P^{Rb}_{td_{j,k}, tw_{j,i}}$:    Percentage of the responsibility of source knock-on delay $tw_{j,i}$ for delay $td_{j,k}$

$tw_{j,i}$:    Knock-on delay of train $j$ on block section $i$

$td_{j,k}$:    Delay of train $j$ on block section $k$

$\chi_{\{l|\, l \in S_{td_{j,k}}\}}$:    Indicator function, if the knock-on delay of train $j$ on block section $l$ $tw_{j,l}$ is the source of the delay of train $j$ on block section $td_{j,k}$, it is equal to 1; otherwise it is equal to 0.

$n_j^{block}$:    Number of block sections along the path of train $j$

$S^{source}_{td_{j,k}}$:    Set of source knock-on delays of delay $td_{j,k}$

With the procedure described above, the relationship between knock-on delays and delays for each train can be quantified. A delayed train may have conflicts with other trains, which results in the knock-on delays of the other trains. As shown in Figure 6-4, three trains Z2, Z4 and Z5 stopped in a station. The trains will depart from the station in the sequence of Z2, Z5 and Z4. Because Z2 will arrive at B6 5 minutes late, the dwell time of Z4 has to be prolonged for 4 minutes. Successively Z5 hindered Z2 additionally 1 minute. As a result Z4 experienced a knock-on delay of 5 minutes on B7.

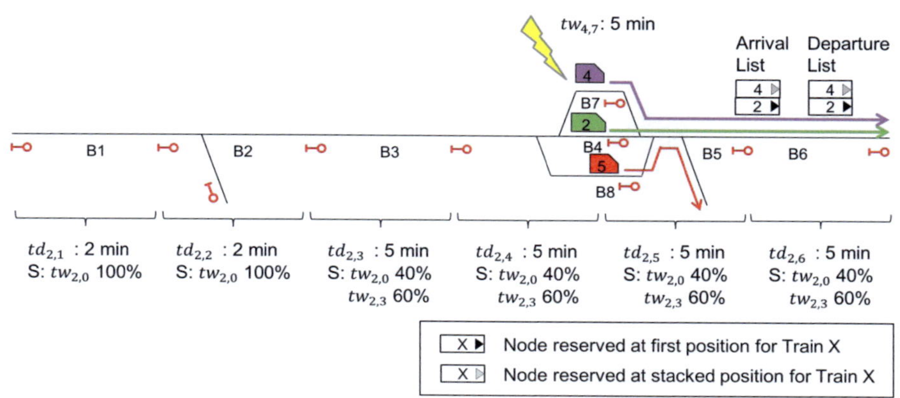


**Figure 6-4: The Delay of a Train Leads to a Knock-on Delay of another Train (source: [Martin and Liang, 2017])**

The percentage of responsibility of each delay for the resulting knock-on delay can be calculated with Formula (6-3): $td_{2,6}$ assumes 80% (4 min / 5 min) responsibility and $td_{5,6}$ assumes 20% (1min / 5 min) responsibility for $tw_{4,7}$. In order to express more clearly, $td_{2,6}$ and $td_{5,6}$ are defined as the source delay of $tw_{4,7}$.

$$P^{Rb}_{tw_{j,i},\, td_{l,k}} = \frac{t_{h_{j,i,l,k}}}{t_{h_{j,i}}} \tag{6-3}$$

Notation used:

$P^{Rb}_{tw_{j,i},\, td_{l,k}}$ : Percentage of responsibility of delay $td_{l,k}$ for knock-on delay $tw_{j,i}$

$t_{h_{j,i,l,k}}$: Time period during which train $j$ on block $i$ had been hindered by train $l$ on block $k$

$t_{h_{j,i}}$: Time period during which train $j$ on block $i$ had been hindered by the other trains

It is difficult to deduce the time period during which Train $j$ on Block $i$ has been hindered by Train $l$ on Block $k$ based upon the protocol of occupation times. Therefore, the time periods of hindrances are counted in the simulation process and outputted as a conflict relationship protocol at the end of the simulation process. Based on the protocol of conflict relationship, the source delays of each knock-on delay and the respective time periods of hindrances will be determined and recorded. In other words, the resulting knock-on delays of each delay and the respective time periods of hindrances are obtained. In the case that a train is hindered by another punctual train, which means the source delay of the knock-on delay does not exist, a fake delay will be created, and the resulting knock-on delays of the fake delay and the respective time periods of hindrances will be also be recorded. A fake delay is defined as the delay of a punctual train that does not actually exist, and it is designed only to aid in the calculation of the percentage of responsibility of delays for knock-on delays. With Formula (6-3), the percentage of responsibility of all delays for their resulting knock-on delays can be calculated.

The relationship between the knock-on delay and the delay of a train, and the influence of the delay of a train on the knock-on delay of another train described above, reveals the basic logic of delay propagation: the knock-on delay of a train leads to the delay of the train itself, and the delay of the train can then further lead to a knock-on delay of another train. Following this basic logic a delay propagation diagram can be drawn for a simulated timetable. An example is shown in Figure 6-5.

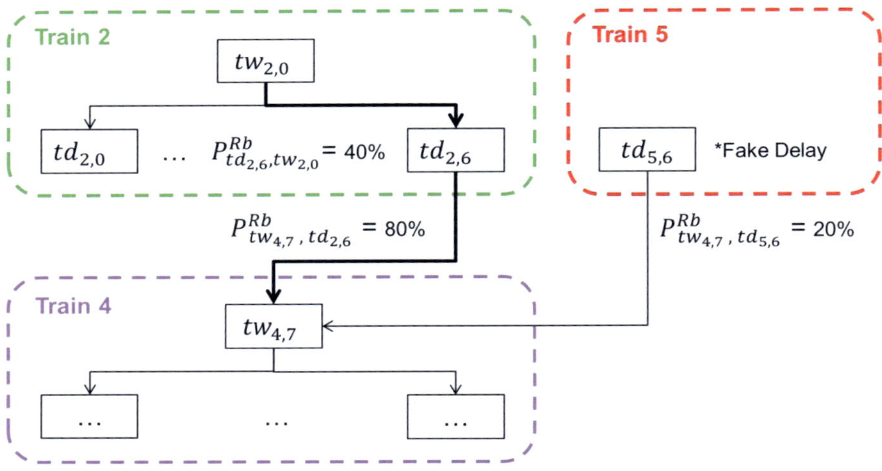


**Figure 6-5: Delay Propagation Diagram (source: [Martin and Liang, 2017])**

Through the delay propagation diagram, the influence of the knock-on delay $tw_{2,0}$ on further conflicts $Inf_{tw_{2,0}}$ can be easily calculated with:

$$Inf_{tw_{2,0}} = P^{Rb}_{td_{2,6},\ tw_{2,0}} \cdot P^{Rb}_{tw_{4,7},\ td_{2,6}} \cdot \widehat{tw}_{4,7}$$

$$+ P^{Rb}_{td_{2,6},\ tw_{2,0}} \cdot P^{Rb}_{tw_{4,7},\ td_{2,6}} \cdot Inf_{tw_{4,7}} \tag{6-4}$$

If $tw_{4,7}$ leads to other knock-on delays on the further path, $Inf_{tw_{4,7}}$ will be calculated with the same method and substituted in Formula (6-4). Otherwise $Inf_{tw_{4,7}}$ is equal to 0. Knock-on delay is weighted in the same manner as in Formula (6-1). The determination of the influence of a knock-on delay on further conflicts is a recursive process, and the recurrence relation follows the same principle as Formula (6-4).

After all knock-on delays are weighted and their influences on further conflicts are calculated, they should be normalized at first.

$$\widehat{tw}_{j,i}{}' = \frac{\widehat{tw}'_{j,i} - \widehat{tw}_{min}}{\widehat{tw}_{max} - \widehat{tw}_{min}} \tag{6-5}$$

$$Inf_{tw_{j,i}}{}' = \frac{Inf_{tw_{j,i}} - Inf_{min}}{Inf_{max} - Inf_{min}} \tag{6-6}$$

Notation used:

$\widehat{tw}_{j,i}{}'$:      Normalized weighted knock-on delay of train $j$ on block section $i$

$\widehat{tw}_{min}$:      Minimum of the weighted knock-on delays

$\widehat{tw}_{max}$:      Maximum of the weighted knock-on delays

$Inf_{tw_{j,i}}{}'$:      Normalized influence of a knock-on delay $tw_{j,i}$ on further conflicts

$Inf_{tw_{j,i}}$:      Influence of a knock-on delay $tw_{j,i}$ on further conflicts

$Inf_{min}$:      Minimum of the influences

$Inf_{max}$:      Maximum of the conflicts

The normalized weighted knock-on delay and their influences on further conflicts will be combined into a comprehensive priority indicator to judge the importance of each knock-on delay. The larger the value of the priority indicator of a knock-on delay is, the more importance the knock-on delay will be. In different application circumstances, such as different system states, the relative importance of the weighted knock-on delay and the influence on further conflicts will vary as well. To that end, a variable representing the relative importance is introduced into the comprehensive priority indicator, which can be adjusted in order to achieve the best performance of the algorithm in different types of application circumstance.

$$Pri_{tw_{j,i}} = C_{tw} \cdot \widehat{tw}_{j,i}' + (1 - C_{tw}) \cdot Inf_{tw_{j,i}}' \tag{6-7}$$

Notation used:

$Pri_{tw_{j,i}}$:     Priority of knock-on delay $tw_{j,i}$

$C_{tw}$:     Relative importance of weighted knock-on delay ($C_{tw} \in [0,1]$)

### 6.1.2     Train Priority Sequence Adjustment

Once a knock-on delay is selected based on the priority rank determined in Section 6.1.1, a series of suitable dispatching actions such as overtaking, passing and replatforming will be carried out to solve conflicts, depending on the characteristics of the knock-on delay. In this dissertation, only three dispatching actions (passing, overtaking and replatforming) are implemented, and only the train priority sequences on loop tracks and open track sections are considered.

Overtaking is intended to exchange the priority sequences between a train and its immediately previous train[21] on an open track section or a loop track, with the two trains having successive movements. Passing is also intended to exchange the priority sequences of a train and its immediately previous train, but with the two trains having opposite movements. The move operation logics of overtaking and passing are the same, so overtaking and passing will be together referred as "overtaking" in the remainder of this report, and both of them are realized through the "swap" move operation. A swap move operation is carried out as follows:

---

[21] The definition of immediately previous train in [Cui, 2010] is used herein: for an open track section T or a loop track LT with more than one train passing, if a train $Z_j$ is not the first train passing T or LT, the immediately previous train $Z_{prev}$ of $Z_j$ is the train passed T or LT before $Z_j$, and there is no other train passing T or LT between $Z_{prev}$ and $Z_j$.

On an open track section T or a loop track LT, the original indexes of priority[22] of two trains $Z_{j1}$ and $Z_{j2}$ are $PI_{j1,T/LT}$ and $PI_{j2,T/LT}$ ($PI_{j1,T/LT} > PI_{j2,T/LT}$). The larger the index of priority of a train is, the earlier the train is scheduled to run through the corresponding loop track or open track section. The indexes of the priority of the two trains on the T or LT will be exchanged by a swap move operation as shown in Figure 6-6, and the indexes of priority of the other trains on the T or LT will remain constant. It is not mandatory that $Z_{j1}$ is the immediately previous train of $Z_{j2}$. It can be seen that a swap move operation is characterized by three attributes: the overtaking train, its immediately previous train and the location where the swap move operation occurred (either an open track section or a loop track).

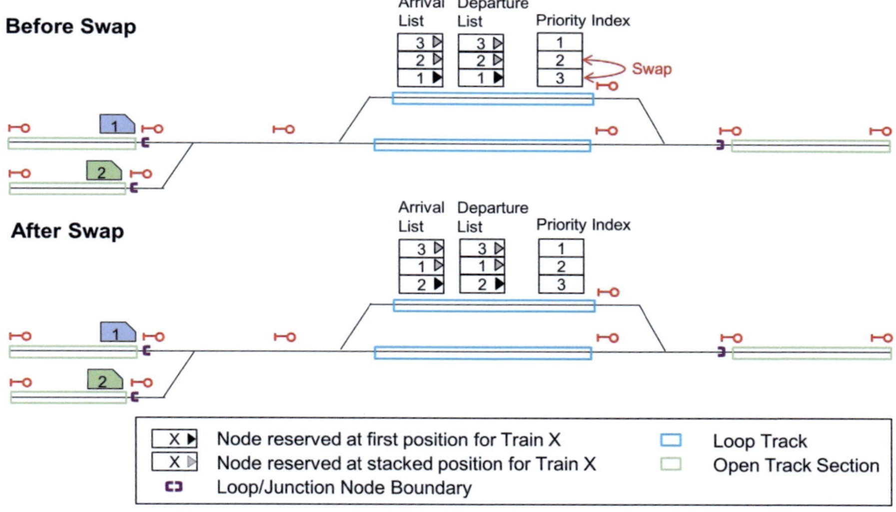


**Figure 6-6: An Example of a Swap Operation on a Loop Track (source: [Martin and Liang, 2017])**

Replatforming is intended to change the position of a train from the original loop track to an alternative loop track, and it is realized by the move operation "insert". An insert move operation is carried out as follows:

---

[22] Index of priority refers to the indexes in both arrival list and departure list on a loop track or an open track section.

On the original loop track LT1, a train $Z_{j1}$ will be removed from the arrival and departure list. Among the other trains on LT1, if $PI_{j2,LT1}$ of a train $Z_{j2}$ is greater than $PI_{j1,LT1}$, it should be updated to $(PI_{j2,LT1} - 1)$. On the alternative loop track LT2, the new index of priority of $Z_{j1}$ is $PI_{j1,LT2}$. Among the pre-existing trains on LT2, if $PI_{j3,LT2}$ of a train $Z_{j3}$ is greater or equal to $PI_{j1,LT2}$, it should be updated to $(PI_{j3,LT2} + 1)$. After all indexes of priority of the pre-existing trains have been updated, $Z_{j1}$ will be inserted into the arrival and departure list according to $PI_{j1,LT2}$. An example of insert move operation is shown in Figure 6-7. The insert move operation is also characterized by three attributes: the to-be-inserted train, the alternative loop track and the immediately previous train of the to-be-inserted train on the alternative loop track. In case the previous train does not exist on the alternative loop track, this attribute should be set as null.

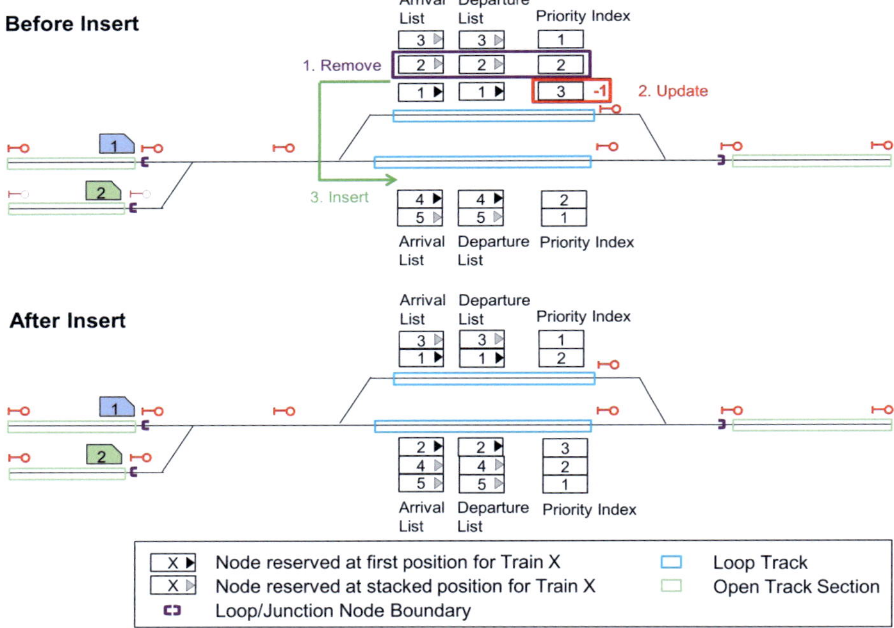


**Figure 6-7: An Example of an Insert Operation (source: [Martin and Liang, 2017])**

In order to solve a knock-on delay properly, a suitable dispatching action will be selected depending on the circumstances of the knock-on delay. As described in Section 6.1.1, a knock-on delay is identified by two attributes: the delayed train and the block section. Firstly, the circumstances will be classified in to four categories depending on the node attribute of the block section (i.e. location of the knock-on delay). The node attribute of the block section is the same as the node attribute of the last basic structure located at the end of the block, and includes four types: loop track, loop non track, junction and open track section. The four categories are defined as follows:

- Category I:  the location of the knock-on delay is a loop track
- Category II:  the location of the knock-on delay is a loop non track
- Category III: the location of the knock-on delay is a junction
- Category IV: the location of the knock-on delay is an open track section

Each category can be divided into several subcategories depending on the characteristics of the further path of the delayed train, and a proper dispatching action will be designated to each subcategory. In order to analyze the further path of the delayed train, a new concept, named macro path, is introduced. The macro path of the train is the sequence of infrastructure nodes along the train path. The further macro path of the delayed train is a partial macro path, which starts from the infrastructure node on which the train was delayed (the node attribute of the block section on which the corresponding knock-on delay occurred) to the last infrastructure node of the whole train path. In the following text, the four categories listed above will be explained individually, and the classification of subcategories for each category will be demonstrated in Figure 6-8, Figure 6-9 and Figure 6-10 respectively.

For categories I and III, the next dispatching relevant infrastructure node in the further macro path of the delayed train will be determined firstly. If the next dispatching relevant node is an open track section, the delayed train will be dispatched to overtake the immediately previous train on the open track section. This case is classified as subcategory I-C or III-C. If the execution of overtaking action has failed, e.g. no previous train, the greedy algorithm will be informed and another knock-on delay will be selected to construct a candidate. If the overtaking action was executed successfully, the relative priority sequences between the immediately previous train and the delayed train on the other part of their common macro path should be analyzed and adjusted to ensure the consistency of the relative priority sequence of these two trains. Violation of priority sequence consistency incurs deadlock problems. The adjustment method for keeping priority sequence consistent will be described at the end of this section.

If the next dispatching relevant node is a loop track (in $Loop_y$) and the delayed train attempts to enter $Loop_y$ (the next block section of the train enters $Loop_y$), replatforming will be executed. This case is classified as subcategory I-A or III-A. To carry out a replatforming action, the following five steps are necessary. Firstly, all possible alternative loop tracks for the delayed train will be selected. Secondly, the occupation time (the same as blocking time) of the delayed train on each alternative loop track will be estimated. Thirdly, the conflicts between the delayed train and the pre-existing trains will be quantified by counting the overlap between the estimated occupation time of the delayed train and the actual occupation times of the pre-existing trains on each alternative track. The alternative track with the minimum number of conflicts is taken as the optimal one for the replatforming action. Finally, the trains on the optimal alternative loop track (including the delayed train) will be sorted by their starts of occupation times; thereby the insert position of the delayed train in the arrival and departure lists of the optimal alternative track can be determined. The replatforming action can be executed through an insert operation as shown in Figure 6-7. After a replatforming action has been completed, the relative priority sequences between the delayed train and each pre-existing train on their common macro path (except the new loop track) could be analyzed and adjusted in order to ensure the priority sequence consistency. However, the involvement of all trains on the new loop track in the priority sequence adjustment could lead to excessive modification of the initial solution; consequently it may result in solution quality deterioration. Furthermore, the computational complexity of adjusting the priority sequence of all trains on a loop track could be extremely high. On one hand it is time consuming, and on the other hand there is a high possibility of generating an infeasible candidate solution. Thus, train priority consistency is not taken into account when a replatforming action is performed. Only if the execution of the replatforming action failed, e.g. no alternative track, the delayed train will be dispatched to overtake its immediately previous train on the original loop track in $Loop_y$.

If the next dispatching relevant infrastructure node is loop track (in $Loop_y$) and the delayed train does not attempt to enter $Loop_y$ at that moment (the next block section of the train does not enter $Loop_y$), the delayed train will be dispatched to overtake the immediately previous train on the loop track. This case is classified as subcategory I-B or III-B. Overtaking on a loop track is executed in the same manner as overtaking on an open track section. In the case of subcategory I-B or III-B, it is capable of providing replatforming possibility in $Loop_y$ for the delayed train. However, as described above, the occupation times of the delayed train on the alternative loop tracks should be estimated in order to select an optimal alternative loop track and determine the insert position. The estimation is based on the assumption that the occupation times of the other trains remain constant, but changes of occupation times will occur as the consequence of dispatching actions. The farther away from $Loop_y$ the delayed train is, the more difficult it is to estimate the occupation time accurately. So a conservative approach is taken in this approach: only if the next block section of the delayed train enters $Loop_y$, replatforming will be considered.

**Category I:knock-on delay on LT ($Loop_y$)**
**Category III: knock-on delay on Junction**

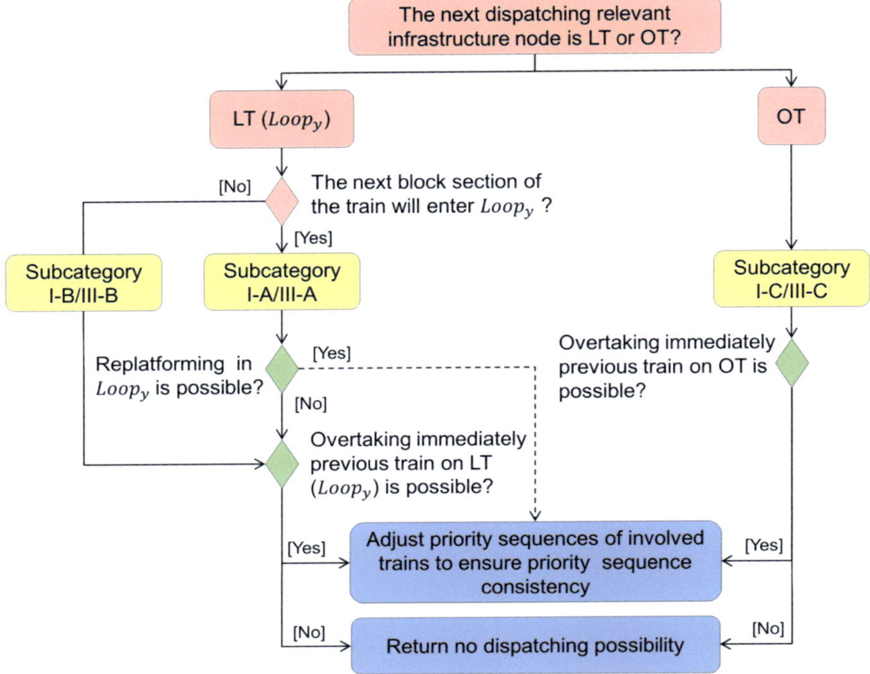


**Figure 6-8: Classification of Subcategories for Category I and Category III of Conflict Circumstance (modified from [Martin and Liang, 2017])**

For Category II, the delayed train is located on a loop non track in $Loop_{y1}$ , and could be approaching or leaving a certain loop track in $Loop_{y1}$. At first, these two cases will be distinguished: if the next loop track of the delayed train is located in $Loop_{y1}$, the train is approaching a loop track in $Loop_{y1}$; otherwise it is leaving. In case of approaching, the delayed train will be dispatched to overtake the immediately previous train on the loop track. This case is classified as subcategory II-A. Similar to the subcategory I-B or III-B described previously, it is also capable of providing the possibility for replatforming in $Loop_{y1}$ for the delayed train in this case. However, the assumption for occupation time estimation is likely inaccurate. The delayed train has already entered the loop at that moment (when the knock-on delay occurs) and the occupation times of the other trains in this loop could be regarded as the interaction results between the delayed train and the other trains. If the delayed train is replatformed, the interactions of the trains will change. In consequence, the occupation times of the trains in the loop will change as well. It is difficult to ensure the accuracy of the estimated occupation times of the delayed train and the corresponding conflicts on the alternative loop tracks. Therefore replatforming is not taken into account for the subcategory II-A in this approach. In case the delayed train is leaving the loop (the next loop track is not located in $Loop_{y1}$), the classification of subcategories and designation of dispatching actions follow exactly the same procedure as that in Category I and Category III.

### Category II: knock-on delay on LNT ($Loop_{y1}$)

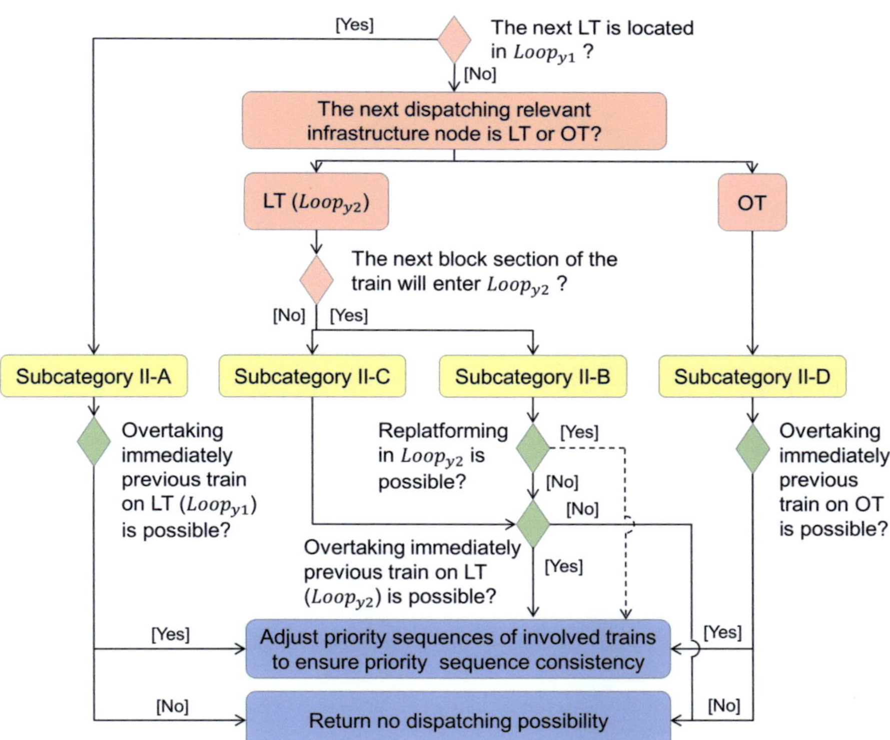


**Figure 6-9: Classification of Subcategories for Category II of Conflict Circumstance**

For Category IV, the delayed train is located on an open track section ($OT_x$). If the current block section of the delayed train (the block section on which the knock-on delay occurred) is not its last block section on $OT_x$, it is indicated that the hindrance is located on $OT_x$. So, the delayed train will be dispatched to overtake its immediately previous train on $OT_x$. On the contrary, if the current block section of the train is the last block section on $OT_x$, the location of the hindrance is certainly beyond $OT_x$. In this case, the next dispatching relevant infrastructure node is concerned with further classification of subcategories and designation of dispatching actions, which follows exactly the same procedure as that in Category I and Category III.

### Category IV: knock-on delay on OT ($OT_{x1}$)

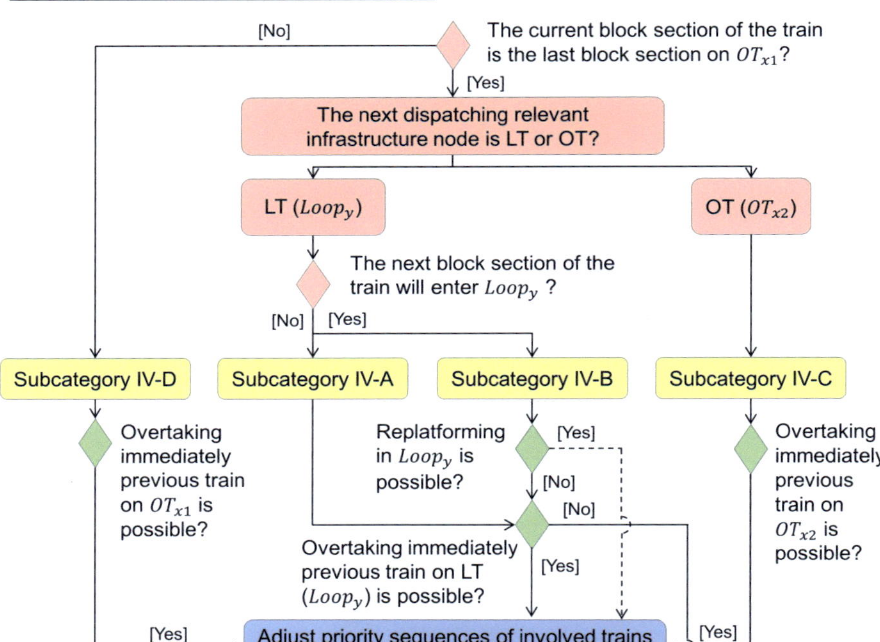


Figure 6-10: Classification of Subcategories for Category IV of Conflict Circumstance

After a dispatching action (overtaking or replatforming) was chosen for a potential candidate solution, the corresponding move operation instance (swap or insert) will be recorded as the move operation attribute of the solution. It is possible that the same dispatching action is chosen for two different selected knock-on delays. For instance, a faster train (denoted by Z1) experienced a series of continuous unscheduled stops on a certain open track section (denoted by OT) because of the hindrance of a previous slower train (denoted by Z2). Several different knock-on delays of Z1 on OT are detected, but the same dispatching action will be chosen – Z1 is dispatched to overtake Z2 on OT. The generated candidate solutions are definitely identical. In order to avoid generating repetitive solutions in one iteration, sensory memory is introduced. The lifecycle of sensory memory is an iteration. It is used to gather the move operation attributes of the existing candidate solutions. When the dispatching action for a new candidate is chosen, it should be compared with the existing ones stored in the sensory memory. In case of duplication the new candidate should be discarded directly, and continue with the next candidate solution.

Once the relative priority sequence between two trains on a certain loop track or open track section is changed successfully through a dispatching action (with consideration of sensory memory), the new relative priority sequence on this loop track or open track section will dictate the relative priority sequence between these two trains throughout the rest of their common partial macro path to ensure the train priority sequence consistency of the newly generated candidate solution. The move operations carried out on the rest of their macro paths can be interpreted as secondary move operations. So only the dispatching action rather than the secondary move operations is taken as the move operation attribute of the candidate solution.

It is supposed that a train $Z_2$ takes precedence over another train $Z_1$ on a certain OT/LT; the index of the OT/LT in the dispatching relevant macro path[23] of $Z_1$ is $p$ (denoted to $N_{Z1,p}^{Dispo}$), and the index of the OT/LT in the dispatching relevant macro path of $Z_2$ is $q$ (denoted to $N_{Z2,q}^{Dispo}$). The adjustment method for the case of $Z_1$ and $Z_2$ with successive movement and that for the case of $Z_1$ and $Z_2$ with opposite movement are different, and they are developed based on the method used in [Cui, 2010].

In case of $Z_1$ and $Z_2$ with successive movement, $Z_1$ should follow $Z_2$ in the forward and backward directions until the two end points of their common dispatching relevant macro path. As shown in Figure 6-11, forward and backward searches are employed to locate the two end points of the common dispatching relevant macro path.

---

[23] The dispatching relevant macro path of a train is the sequence of dispatching relevant nodes (loop tracks and open track sections) along its train path.

Forward search is taken as an example (backward search follows the same principle). Firstly, the next dispatching relevant nodes $N_{Z1,p+1}^{Dispo}$ and $N_{Z2,q+1}^{Dispo}$ are determined for $Z_1$ and $Z_2$. There are three possibilities:

- $N_{Z1,p+1}^{Dispo}$ and $N_{Z2,q+1}^{Dispo}$ are not the same or they are the same $OT_x$ composed of free resources. The end point in the forward direction is reached and the forward search terminates.

- $N_{Z1,p+1}^{Dispo}$ and $N_{Z2,q+1}^{Dispo}$ are the same $LT_z$ (in $Loop_y$). $Z_1$ and $Z_2$ will be attempted to be replatformed in $Loop_y$. If one of them is replatformed successfully, the forward search terminates; otherwise the priority sequence of $Z_1$ and $Z_2$ should be swapped on $LT_z$, and the forward search continues.

- $N_{Z1,p+1}^{Dispo}$ and $N_{Z2,q+1}^{Dispo}$ are the same $OT_x$ composed of non-free resources. The priority sequence of $Z_1$ and $Z_2$ should be swapped on $OT_x$ , and the forward search continues.

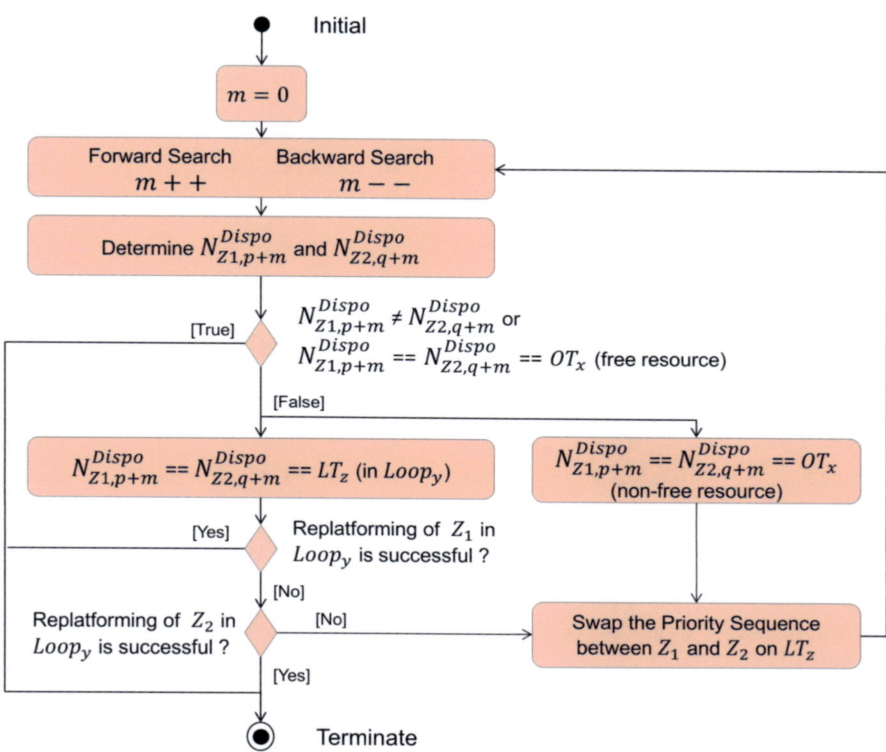


**Figure 6-11: Priority Sequence Adjustment for Two Trains with Successive Movement (Forward Search and Backward Search) and for Two Trains with Opposite Movement (Forward Search) (source: [Martin and Liang, 2017])**

In case of $Z_1$ and $Z_2$ with opposite movement, the priority sequence adjustment in the forward direction follows the same principle of the forward search for two trains with successive movement. The forward direction particularly refers to the proceeding direction of $Z_2$ (the overtaking train rather than the overtaken train). Because the running directions of these two trains are opposite, the method of determining the next dispatching relevant node for $Z_1$ should be adjusted as shown in Figure 6-12.

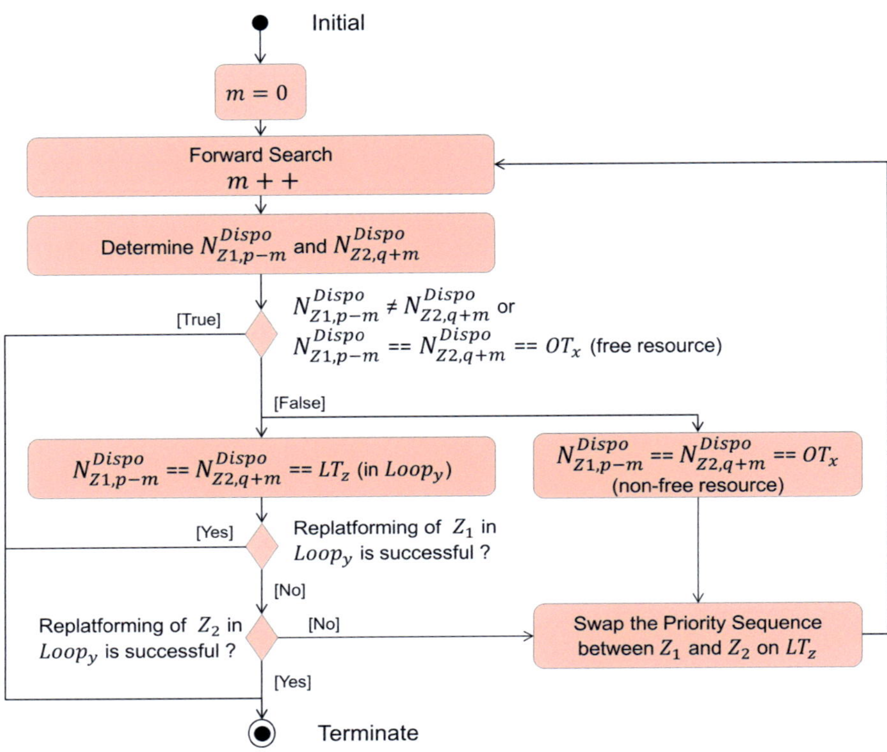



**Figure 6-12: Priority Sequence Adjustment for Two Trains with Opposite Movement (Forward Search)**

For the backward search, it is not sufficient to guarantee the priority sequence consistency of $Z_1$ and $Z_2$ only by the adjustment of their common dispatching relevant macro path. An example is shown in the subgraph a) of Figure 6-13: $Z_2$ overtook $Z_1$ on an open track section OT1, and the next loop is found through the backward search. Following the principle in Figure 6-11, $Z_1$ will be replatformed (from LT2 to LT1) in this loop, and then the backward search terminates. However, the generated solution fell into a deadlock situation as shown in the subgraph b) of Figure 6-13: OT1 is reserved at the first position for $Z_2$ and at the stacked position for $Z_1$, which indicates that $Z_1$ should wait until $Z_2$ has completely passed through OT1;at the other end, $Z_2$ is waiting for $Z_1$ for the same reason. Obviously a circular wait situation occurred in this case, which resulted in a deadlock solution. This kind of deadlock will not occur in reality, but is likely to occur in a simulation environment due to the inconsistency of train priority sequences. It can be seen that it is necessary to extend the backward search beyond the end of the common dispatching relevant macro path. The terminate specification of the backward search for two trains with opposite movement are summarized into four indicators, if one of them is fulfilled, the backward search will terminate.

-   $N_{Z1,p+m}^{Dispo}$ is a loop track (or an open track section), and $N_{Z2,q-m}^{Dispo}$ is an open track section (or a loop track).

-   $N_{Z1,p+m}^{Dispo}$ and $N_{Z2,q-m}^{Dispo}$ are loop tracks belonging to two different loops.

-   $N_{Z1,p+m}^{Dispo}$ and $N_{Z2,q-m}^{Dispo}$ are two different open track sections.

-   $N_{Z1,p+m}^{Dispo}$ and $N_{Z2,q-m}^{Dispo}$ are the same open track section composed of free resource.

With the new terminate specification, the initial solution in Figure 6-13 is readjusted, the priority sequence of $Z_1$ and $Z_2$ is swapped in the backward direction until the two trains reach two different open track sections. The new generated solution is deadlock-free as shown in the subgraph c) of Figure 6-13.

## Backward Search (Opposite Movement) →

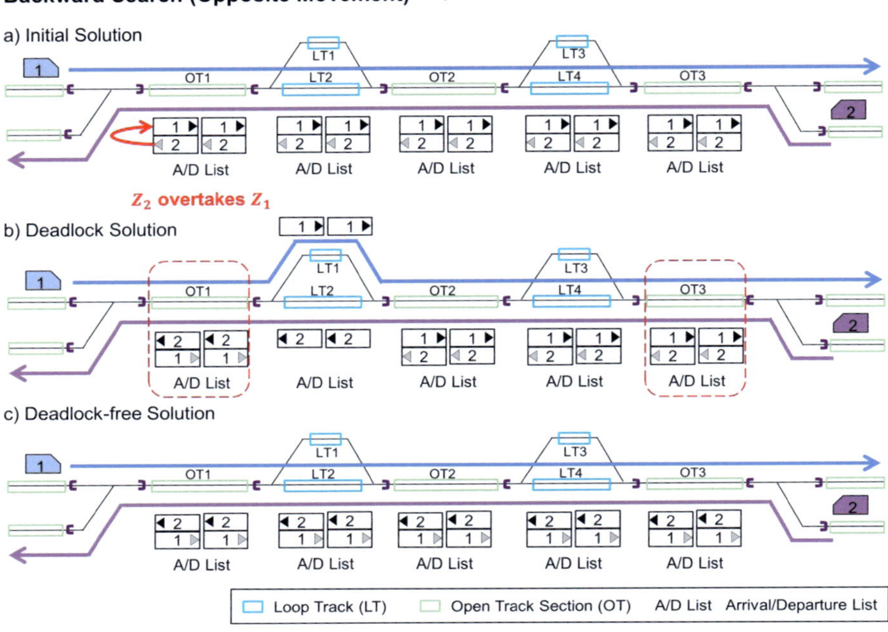


**Figure 6-13: Improper/Proper End of Backward Search for Two Trains with Opposite Movement (source: [Martin and Liang, 2017])**

The procedure of priority sequence adjustment in the backward direction for two trains with opposite movement is shown in Figure 6-14. Similar to the backward search for two trains with successive movement, the next dispatching relevant nodes $N_{Z1,p+1}^{Dispo}$ and $N_{Z2,q-1}^{Dispo}$ in the backward direction will be firstly determined for $Z_1$ and $Z_2$. If the terminate specification is fulfilled, the backward search terminates. Otherwise, the following possibilities exist:

- $N_{Z1,p+1}^{Dispo}$ and $N_{Z2,q-1}^{Dispo}$ are the same loop track ($LT_z$) or open track section ($OT_x$). The priority sequence between $Z_1$ and $Z_2$ will be swapped on $LT_z/OT_x$.

- $N_{Z1,p+1}^{Dispo}$ and $N_{Z2,q-1}^{Dispo}$ are two different loop tracks belonging to the same loop. No action to be taken, backward search continues.

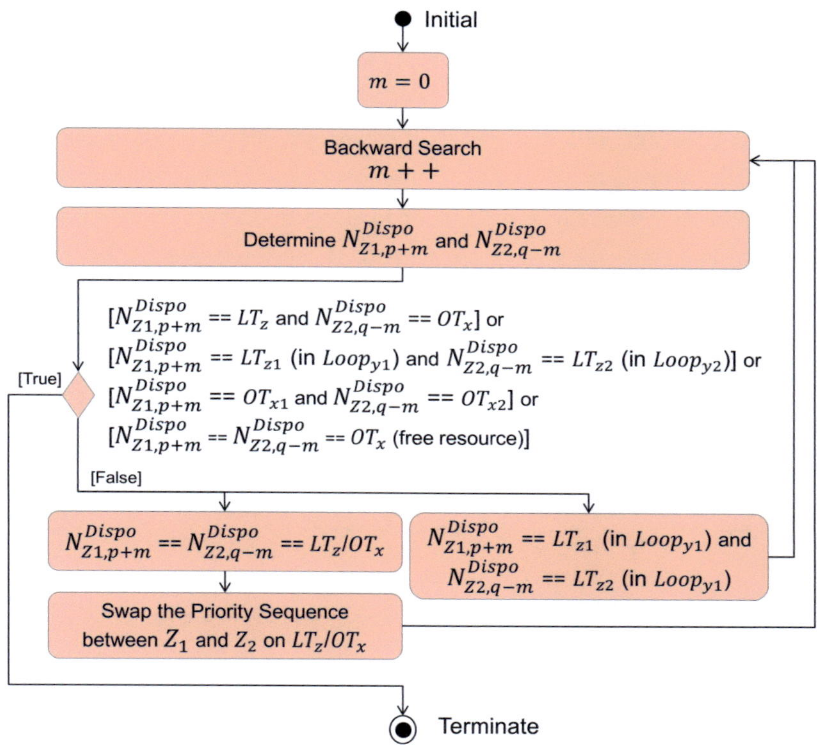


**Figure 6-14: Priority Sequence Adjustment for Two Trains with Opposite Movement (Backward Search) (source: [Martin and Liang, 2017])**

### 6.1.3　Deadlock-free Test

The candidate solutions generated in Section 6.1.2 may be infeasible because of deadlock problems. Before the simulation of a candidate solution, a deadlock-free test will be executed to analyze the feasibility of the given candidate solution. A candidate solution that cannot pass the deadlock-free test will be abandoned.

Deadlock detection is irrelevant to running time calculations, and only concerns the priority sequences on all loop tracks and open track sections. The given priority sequences will be analyzed following the process illustrated in Figure 6-15. In the first step, the macro paths of all trains, as well as the arrival and departure lists on all loop tracks and open track sections will be initialized, taking the initial position of a train as the first infrastructure node along its macro path. In the second step, the state of each train will be analyzed. Only if a train has already reached the last infrastructure node along its macro path, which is definitely a free resource, is the train movement regarded as terminated; otherwise the train movement is active. If all trains can terminate their movement eventually, the solution is denoted as deadlock-free. In case some trains have not yet terminated their movement, the positions of these active trains will be attempted to be updated in the third step. The position of a train will be updated to the next dispatching relevant node along its macro path in one step if the train priority sequence constraints are fulfilled. In the fourth step, it will be judged whether a circular wait situation has occurred (none of the active trains moved forward in one step). If so, the solution is denoted as a deadlock; otherwise, the second step of this process will be repeated. This process (except initialization) will be iterated continuously until either a circular wait situation occurs or all train movements are terminated successfully.

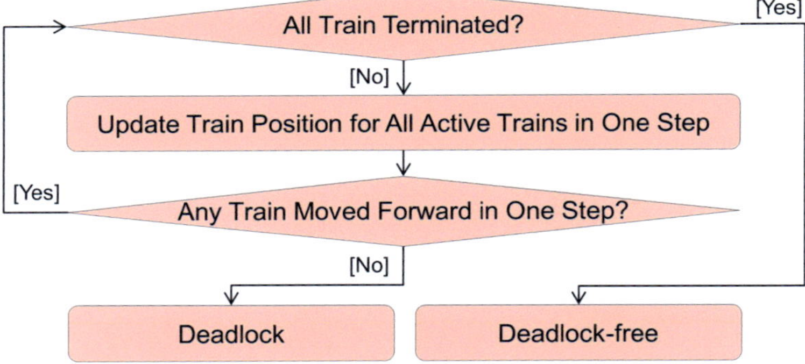


**Figure 6-15: The Process of Deadlock-free Test**

To update the position of a train, the priority sequence constraints on relevant infrastructure nodes should be obeyed. The logic of the priority sequence control has been clearly described in Chapter 5, which regulates the train movements on the basis of the applied signaling systems in the simulation model. However, the train movements in the deadlock-free test are only regulated by the priority sequence constraints in order to improve the computational efficiency of the algorithm. So the logic of the priority sequence control in Chapter 5 was expanded with additional constraints, and the new mechanism of priority sequence control is shown in Figure 6-16. In the following text the priority sequence control for deadlock-free test will be explained step by step.

**Step 1:** If the train is on its initial position along its macro path (a free resource[24]), the train movement is regulated by the priority sequence on the next loop track (LT) or open track section (OT). In this case, proceed directly to Step 4; otherwise continue to the following step.

**Step 2:** Only if the current OT/LT is reserved at the first position for this train in the departure list, the train is allowed to leave the current OT/LT and request the next OT/LT. This constraint is additionally integrated due to the neglect of signaling systems. An example is shown in Figure 6-16: when signaling systems exist, Train 2 cannot leave OT1 at that moment, since it cannot obtain the next block section occupied by Train 1; without signaling systems, the sequence of leaving has to be restricted by the departure list. If the train is allowed to leave the current node, continue to the next step; otherwise the position of the train should not be updated (train movement is not allowed).

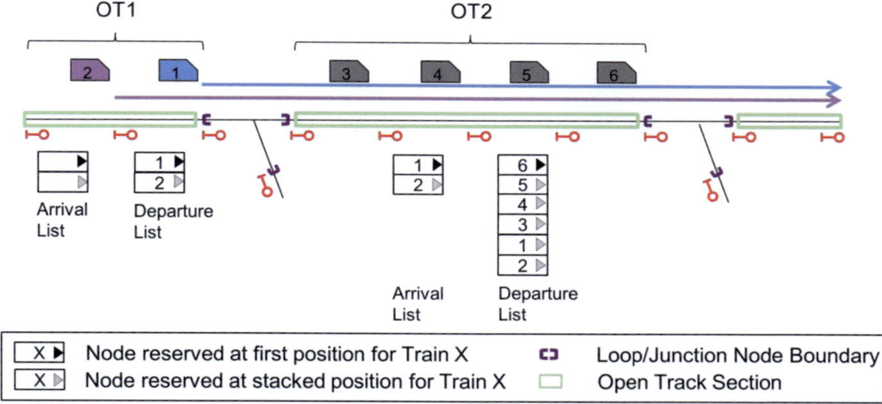


**Figure 6-16: An Example of Additional Priority Sequence Constraints in Deadlock-free Test**

**Step 3:** In case that the next OT/LT of the train is a free resource, which is conclusively the end of the macro path, the train movement will be directly permitted. A train movement includes three actions: firstly the train will be removed from the departure list of its current OT/LT if the current OT/LT is not a free resource; secondly, the train will be removed from the arrival list of the next OT/LT if the next OT/LT is not a free resource; finally, the current train position will be updated to the next OT/LT. On the

---

[24] There is no priority sequence control on open track sections that consist of free resources.

contrary, if the next OT/LT is not free resource, the priority sequence constraints on the next OT/LT should be checked in the next step.

**Step 4:** If the next OT/LT is not reserved at the first position for the train in the arrival list, the train movement is not allowed; otherwise proceed to the next step.

**Step 5:** If the next OT/LT is reserved at the first position for the train in the departure list, the train movement will be allowed and the corresponding three actions will be carried out; otherwise proceed to the next step.

**Step 6:** If the running direction of this train and its immediately previous train in the departure list is not the same, the train movement should not be allowed to avoid deadlock problems; otherwise proceed to the next step.

**Step 7:** The difference between the lengths of departure and arrival list is the number of trains concurrently occupying an OT/LT at a certain movement, and the maximum capacity of an OT/LT is defined as the maximum number of trains that can concurrently occupy it. The maximum capacity constraint is also additionally integrated due to the neglect of signaling systems. As shown in Figure 6-16, Train 1 should not be allowed to enter OT2 at the current moment, since the capacity of OT2 is exhausted. So, if the number of existing trains on the next OT/LT does not exceed the maximum capacity, the train movement is allowed; otherwise the position of the train should not be updated (train movement is not allowed).

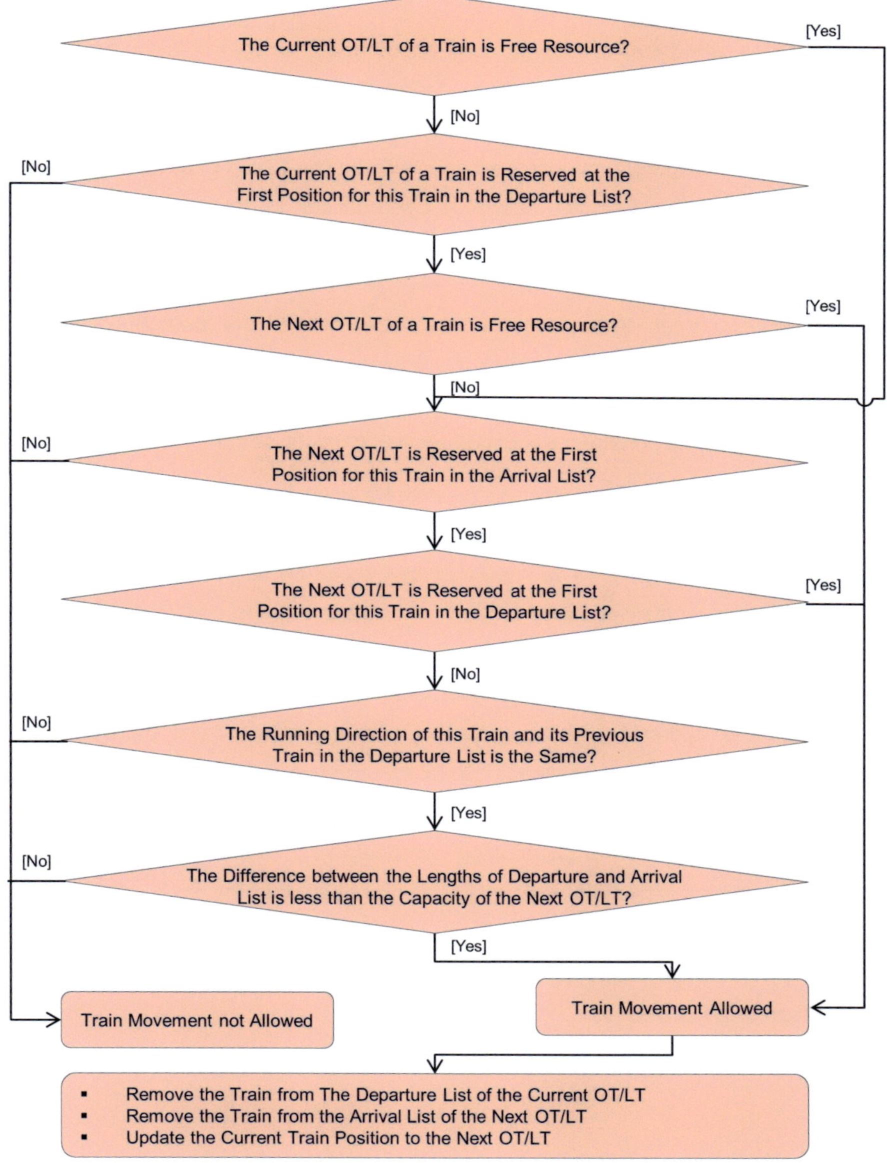



**Figure 6-17: Stepwise Update of the Position of a Train (OT: Open Track, LT: Loop Track)**

### 6.1.4 Candidate List Strategy

In tabu search, the intensification strategy encourages the search to concentrate on good regions and good solution features, while the diversification strategy encourages the search to exploring unvisited regions [Glover et al., 2007]. Both are equally important for achieving global optimization. However, the timing of switching between intensification and diversification search is often difficult to determine. In [Liu et al., 2014], an adaptive search strategy is proposed to solve the travelling salesman problem. Both intensification and diversification elements are included in the candidate list, and their numbers are dynamically adjusted according to the changes of solution quality during the search process. With this adaptive search strategy, the intensification and diversification searches are well-balanced. Therefore, the adaptive search strategy is adopted to construct the candidate list in this dissertation.

Following LIU's adaptive search strategy stated above, the candidate list of the implemented tabu search algorithm consists of intensification elements and diversification elements (Figure 6-18). The higher ranked knock-on delays will be selected in sequence based on their priorities to generate a certain amount of intensification elements, while the left lower ranked knock-on delays will be randomly selected to generate a certain amount of diversification elements. The numbers of the two types of elements (denoted by $L_{Int}$ and $L_{Div}$ respectively) are initialized with the same starting value (i.e. ½ candidate list length[25]). In the iterative process, the best solution found in the current iteration will be compared with the best solution found in the previous iteration. If the value of dispatching objective function is reduced in the current iteration, $L_{Int}$ will be increased by one ($L_{Div}$ will be decreased by one); otherwise $L_{Int}$ will be decreased by one ($L_{Div}$ will be increased by one). In order to ensure that both types of elements are included in the candidate list, $L_{Int}$ and $L_{Div}$ should be always greater than or equal to one.

---

[25] The maximum length of the candidate list should be set by the user before the dispatching optimization algorithm starts.

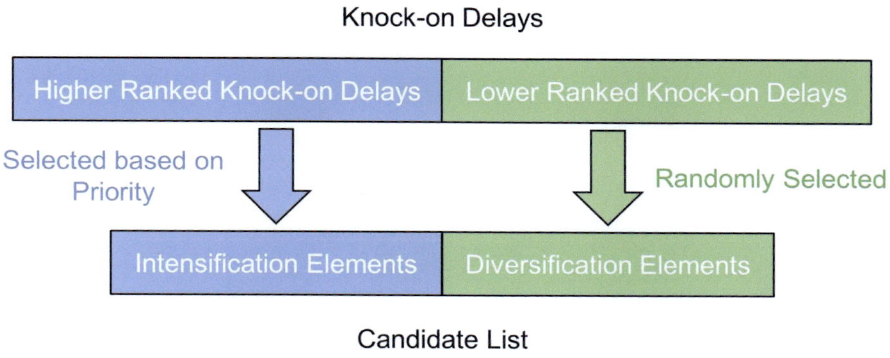

**Figure 6-18: Construction of the Candidate List with Intensification and Diversification Elements**

The advantage of this adaptive search strategy is very intuitive. When the solution quality is improved (value of dispatching objective function is reduced), it implies that an attractive region of the search space may be achieved. So it is meaningful to reinforce intensification search by increasing the number of intensification elements. On the contrary, when the solution quality is deteriorated, it is more meaningful to explore the unvisited regions of the search space through the randomly selected diversification elements. In addition, diversification elements can also help the optimization algorithm escape from local optima.

## 6.2  Evaluation of Candidate Solutions

For the evaluation of candidate solutions, the dispatching objective functions in [Martin, 1995] are used. Two dispatching objective functions are defined: punctuality and fluency of operation. Under different dispatching conditions, the importance of objectives differs. In [Martin, 1995], viscosity is introduced to evaluate dispatching conditions. Viscosity is defined as the total knock-on delay of all the conflicted trains divided by the number of basic structures of the dispatched network. Punctuality is the primary objective when viscosity is small, and fluency of operation is the primary objective when viscosity is large. By introducing viscosity, these two objective functions are combined into a general objective function (Formula (6-8)).

$$\sum_{j=1}^{n_{ges}} \left( \frac{C_j + Zfl}{1 + Zfl} \cdot \left( \sum_{i=1}^{zj} tw_{j,i} \right) + \frac{1}{1 + Zfl} \cdot \max\{t_{Pein\,j} + t_{Pur\,j} - \sum_{i=1}^{zj} t_{R\,j,i}; 0\} \right)$$ (6-8)

$$\Rightarrow MIN$$

Notations used:

$tw_{j,i}$:  Knock-on delay of train $j$ on block section $i$

$t_{Pein\,j}$:  Original delay of train $j$

$t_{Pur\,j}$:  Initial delay of train $j$

$t_{R\,j,i}$:  Recovery time of train $j$ on block section $i$

$C_j$:  Constant for weighting knock-on delays of train $j$

$n_{ges}$:  Total number of trains

$zj$:  Amount of block sections along the path of train $j$

$Zfl$:  Viscosity of a certain dispatching condition

Timetables with stochastic deviations generated by the software PULEIV ([Martin et al. 2008a; Martin et al. 2008d; Martin et al. 2008c]) are going to be optimized in Section 6.4, in which recovery times are not included. Therefore, the second part of the objective function – delay reduction $max\{...\}$ - is constant, which do not have influence on the minimization of the value of dispatching objective function. So Formula (6-8) is simplified into the following form:

$$\sum_{j=1}^{n_{ges}} \left( \frac{C_j + Zfl}{1 + Zfl} \cdot \left( \sum_{i=1}^{zj} tw_{j,i} \right) \right) \Rightarrow MIN$$ (6-9)

The actual occupation and releasing times of basic structures will be recorded and outputted as a protocol at the end of a simulation. The scheduled occupation and releasing times should be ready beforehand. By comparing the scheduled and actual data, the knock-on delays can be determined.

By definition the calculation method of viscosity is as follows:

$$Zfl = \frac{\sum_{j=1}^{n_{ges}} \sum_{i=1}^{zj} tw_{j,i}}{a_{BS}} \tag{6-10}$$

Notation used:

$Zfl$:  Viscosity of a certain dispatching condition

$a_{BS}$:  Amount of basic structures in the investigation area

To determine the constant for weighting knock-on delay, two types of indicators are used in [Martin, 1995] and [Cui, 2010]: class-oriented and train oriented indicators. Class-oriented indicators include the type of trains, passing or stopping criteria and punctuality criteria. Based on these indicators, a constant for weighting knock-on delays (denoted by $C^c$) will be calculated for a group of trains with common characteristics. The train oriented indicator refers to the possibility of reduction of delay in further movement[26], based on which another constant (denoted by $C^t$) will be calculated for each specific concerned train. For the detailed calculation method, please refer to [Martin, 1995] and [Cui, 2010]. By adding the constant $C_j^t$ to the respective constant $C_k^c$, the final constant $C_j$ for weighting a knock-on delay for a train $j$ is obtained [Cui, 2010].

$$C_j = C_k^c + \alpha \cdot C_j^t \tag{6-11}$$

The empirical constant value $\alpha$ is used to normalize $C_j$ to keep $0 \leq C_j \leq \max(C_k^c)$.

## 6.3    Tabu List and Terminate Specification

The primary goal of a tabu list is to prevent the search process from being trapped in local optima. The move operation attributes in the recently visited solutions are recorded in the tabu list and become tabu-active during their tabu tenures. If the move operation attribute of a candidate solution is tabu-active, it should be forbidden to be visited without regard to the aspiration criterion.

---

[26] As stated above, recovery times are excluded in the timetables with stochastic deviations generated by the software PULEIV. Therefore, in the simulation experiments to be carried out in Section 6.4, the possibility of reduction of delay in further movement is equal to zero.

As stated in Section 6.1.2, two types of move operation (swap and insert) are utilized in this approach. Accordingly, two types of tabu list – tabu swap list and tabu insert list - are used simultaneously to record the tabu-active swap move operation and insert move operation respectively. Furthermore, two types of tabu tenures are defined as follows, and can be set to different values.

- Tabu swap tenure: number of iterations during which a swap move operation is tabu-active

- Tabu insert tenure: number of iterations during which an insert move operation is tabu-active

When tabu restrictions become unreasonable in some special cases, an aspiration criterion is necessary to revoke the move operation attribute's tabu-active status. Two aspiration criteria in [Glover and Laguna, 1993] are implemented within this approach, which also have been widely used in other researches.

- Aspiration criterion 1: if the quality of a tabu-active candidate solution is better than the best solution found so far, revoke its tabu classification.

- Aspiration criterion 2: if the move operation attributes of all candidate solutions are tabu and aspiration criterion 1 is not fulfilled, select the best candidate solution in the current iteration.

The selection of the best neighbor in one iteration is restricted by both tabu lists and aspiration criteria, and the selection procedure is summarized in Figure 6-19. The best admissible candidate solution will be used to replace the initial solution. If the terminate specification is not fulfilled in the current iteration, then continue with the next iteration.

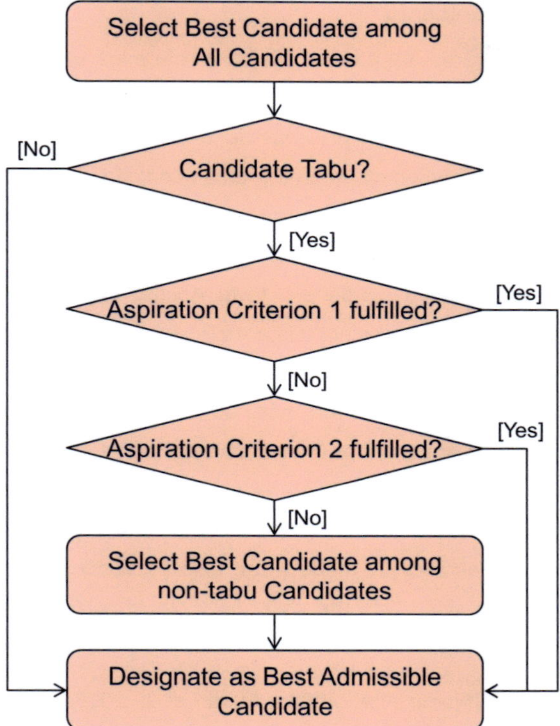


**Figure 6-19: Selection of the Best Admissible Candidate for Tabu Search**

In this approach, total computation time is implemented as the termination criteria: after a certain computation time the optimization algorithm must terminate. The total computation time available for dispatching tasks should be dynamically set depending on on-line factors, such as the time span of prediction, the number of trains involved in conflicts, the scope of delay propagation and the urgency of solving conflicts. For instance, the computation time for a conflict going to occur within the next ten minutes should be much shorter than that for a conflict within the next one hour. The proper computation time for each special case can be determined in advance based on historical operation data (e.g. through knowledge-based expert systems). For on-lines applications, the proper computation can be quickly set according to matching historical records. The determination of computation time will not be covered in this dissertation, and assumed values are used to test the approach.

## 6.4    Simulation Experiments

The simulation experiments are carried out on a reference example. The sketch of the complete infrastructure network of the reference example is presented in the first subgraph of Figure 6-20. Through previous simulation experiments, it is found that conflicts between trains (especially merging and opposing conflicts between trains) mainly concentrated in the area between the stations AHX, BS, EN, LBC. Therefore, this area is chosen as the investigated area as shown in the first subgraph, and its simulation is carried out on the microscopic level. Accordingly, this area is zoomed in, and the microscopic subnetwork within this area is presented in the second subgraph of Figure 6-20. Based on the multi-scale concept, the other areas around the investigated area should be simulated on the more efficient mesoscopic and macroscopic levels. However, the mesoscopic and macroscopic models have not been implemented in the simulation software yet, so, dispatching optimizations are only carried out on the microscopic level in the investigated area in this dissertation. In principle, the mesoscopic and macroscopic models can also be successfully implemented in the simulation software, since they are essentially two simplified forms of the microscopic model.

The software PULEIV ([Martin et al. 2008a; Martin et al. 2008d; Martin et al. 2008c]) was used to generate timetables with stochastic deviations based on a basic operating program. The time interval of each generated timetable is 6 hours, which is composed of three parts (2 hours each): preheating time, investigated time period and cool down time [Chu, 2014]. The investigated time period represents the rush hour in the investigated area (i.e. the reference example). Only the traffic situation in the investigated time period will be optimized in this example.

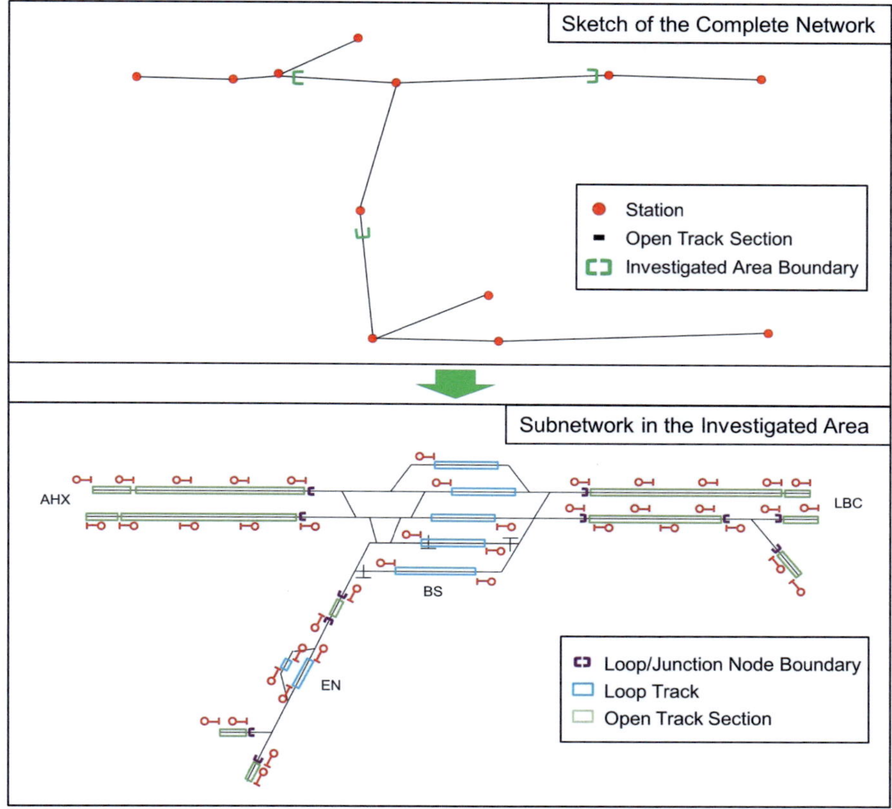


**Figure 6-20: Infrastructure Network of the Reference Example**

The structure of the basic operating program is shown in Table 6-1. Three types of trains are defined: long distance passenger trains (FRZ), short distance passenger trains (NRZ) and freight trains (GV). Four running directions are included: from Station AHX to LBC, from Station LBC to AHX, from Station EN to LBC and from Station LBC to EN. The recommended area of traffic flow (OLB) determined through capacity research is 15-22 Z/h (trains per hour), which can be interpreted as the optimal traffic load for this example (for the detailed calculation method of OLB it is referred to [Martin, 2014]). Several timetables within this range are randomly generated with PULEIV, and used to analyze the performance of the dispatching optimization algorithm.

| Train Path Group | Base Load | Train Path Group | Base Load |
|---|---|---|---|
| FRZ/AHX-LBC | 1.0 Z/h | NRZ/AHX-LBC | 2.0 Z/h |
| FRZ/LBC-AHX | 1.0 Z/h | NRZ/EN-LBC | 1.0 Z/h |
| GV/LBC-EN with Stop | 0.5 Z/h | NRZ/LBC-AHX | 2.0 Z/h |
| GV/EN-LBC with Stop | 0.5 Z/h | NRZ/LBC-EN | 1.0 Z/h |

**Table 6-1: Basic Operating Program on the Reference Example**

Before the optimization process starts, the parameters of optimization algorithm should be setup (Table 6-2). The weights of knock-on delay for FRZ, NRZ and GV is calculated with the method developed in [Martin, 1995] and [Cui, 2010]. The detailed calculation procedure can be found at Section 5.1.3 in [Cui, 2010]. The parameters including tabu swap tenure, tabu insert tenure, candidate list length and relative importance of weighted knock-on delay are set to a set of fixed empirical values. These parameters can be case-specifically optimized based on the historical operational data, and the optimal value of a parameter may vary under different system states. In [Martin and Liang, 2017] a system state classification method is developed based on the theory of capacity research of railway operation, which is intended to improve the performance of dispatching algorithms by providing of adjusting the settings of dispatching parameters in different system states. Optimization of dispatching parameters will not be covered in this dissertation, for more details it is referred to [Martin and Liang, 2017]. For on-line dispatching, the dispatching time horizon is limited, such as the next 30 minutes, because train runs are likely to be disturbed in a long operations planning horizon [Jacobs, 2008]. In most cases, the identified conflicts have to be resolved in a short time. However, on one hand, timetables of 6 hours have to be simulated as a whole to create traffic conditions during the rush hour; on the other hand, the dispatching time horizon is 2 hours in this example. So a relative long maximum computation time (6 minutes) is chosen to show the effects of the model clearly.

| | | | |
|---|---|---|---|
| Tabu Swap Tenure | 3 | Weight of Knock-on delay for FRZ ($C_j$) | 1.80 |
| Tabu Insert Tenure | 3 | Weight of Knock-on delay for NRZ ($C_j$) | 1.33 |
| Candidate List Length | 5 | Weight of Knock-on delay for GV ($C_j$) | 0.83 |
| Terminate Specification | 6 min | Relative Importance of weighted Knock-on Delay ($C_{tw}$) | 0.7 |

**Table 6-2: Settings of the Basic Parameters for the Dispatching Optimization Algorithm**

The optimization model runs on a PC equipped with a processor Intel Core i5-4670 (3.40GHz), 8G RAM and Windows 7 operating system. The performance of the dispatching optimization algorithm on two test cases is presented in Figure 6-21 and Figure 6-22, and more results are attached in Appendix II. In order to make the experimental results more intuitive and easier to be understood, besides total weighted knock-on delay, total unweighted knock-on delay is also presented in the figures. The dispatching optimization algorithm has produced promising results.

- Compared to the initial solution, the total weighted knock-on delay can be significantly reduced. For instance, the total weighted knock-on delay is reduced by 27.4% in the Test Case 1 and by 24.7 % in the Test Case 2 as shown in Figure 6-21 and Figure 6-22. Theoretically, there is also the possibility that the initial solution generated based on FCFS principle already provides good results, and therefore the improvement in solution quality is limited. Because the FCFS dispatching principle is also an integral part of the dispatching optimization module. So in either case, satisfactory results can be provided.

- With implementation of the adaptive search strategy, intensification and diversification have been well balanced. Under the guidance of the intensification search strategy, the optimization algorithm is capable to find suboptimal or optimal solutions in a short time. For instance, for both Test Case 1 and Test Case 2, a suboptimal solution (i.e. Solution S1 and Solution S4 marked in Figure 6-21 and Figure 6-22) was found in less than 50 seconds. Moreover, once a good search direction is found, the intensification strategy can guide the optimization algorithm to improve solution quality very sharply (the trend are indicated with purple dashed lines in Figure 6-21 and Figure 6-22). With the help of the tabu list and diversification search strategy (i.e. randomly selected can-didates), the search process is capable of escaping from local optima (see red dashed lines in Figure 6-21 and 6-22), and the search scope of the optimization algorithm is broadened as well. With a broadened search scope, the possibility of finding a better solution is increased. For both Test Case 1 and Test Case 2, the best solutions (i.e. S3 and S7) are found after escaping from the local optimal solutions (i.e. S2 and S6). Compared to typical local search algorithms (e.g. greedy algorithm), the advantage of a tabu search-based optimization algorithm is that it has a higher possibility to find a better solution in a limited time span.

— Last but not least, it can be seen that from the illustration of the eight test cases: the relationship between total weighted knock-on delay and total un-weighted knock-on delay could be approximated as positive correlation, and the difference between them is not significant, which means knock-on delays are weighted in a proper manner, without deteriorating knock-on delays of lower ranked trains severely.

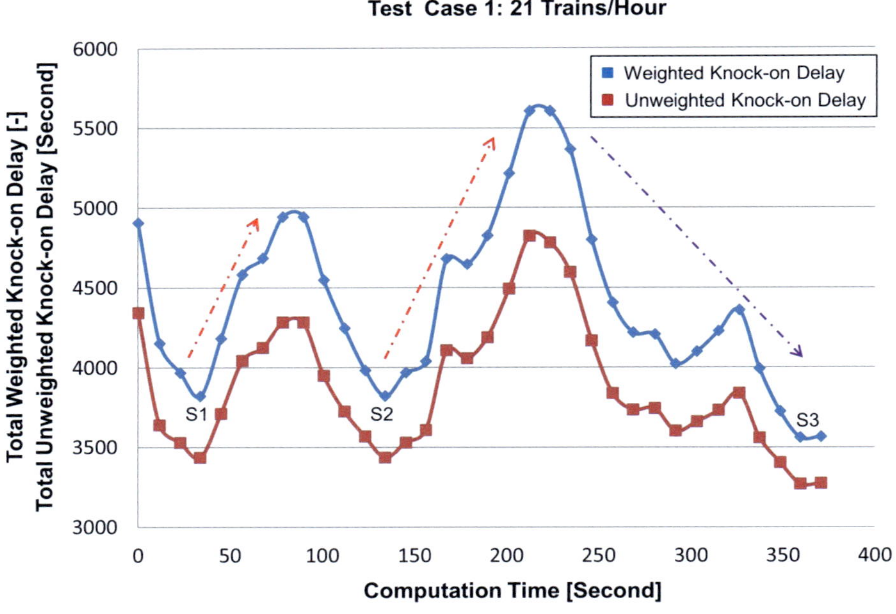



**Figure 6-21: Dispatching Optimization Algorithm – Test Case 1**

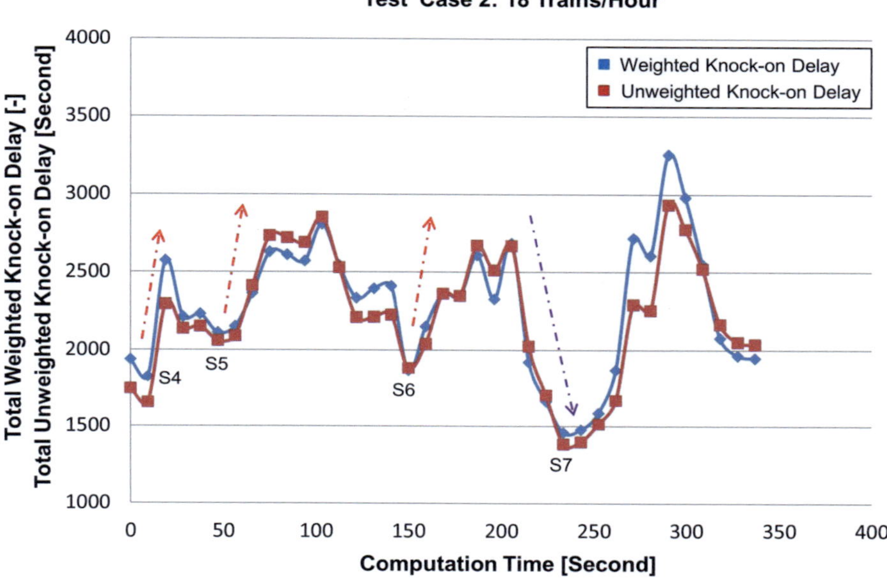


**Figure 6-22: Dispatching Optimization Algorithm – Test Case 2**

In order to compare the experimental results comprehensively, the final results of all eight cases are summarized in Table 6-3. Because total weighted knock-on delay and total unweighted knock-on delay are approximately linearly correlated, only the data on total weighted knock-on delay is shown in Table 6-3. The eight test cases covered a relatively wide range of traffic load, which is ranging from 15.5 to 22 Trains/Hour. Moreover, the qualities of the initial solutions also vary significantly (the total weighted knock-on delays range from 1486 to 12468). Even through the characteristics of the test cases are very different; the dispatching optimization algorithm has successfully reduced the total weighted knock-on delays very significantly on all test cases. On average the total weighted knock-on delay is reduced by 48%. For the severely disturbed situations (i.e. Test Case 3 and Test Case 4), the reduction of total weighted knock-on delay even reached 70%. Through the above simulation experiments, the effectiveness of the dispatching optimization algorithm developed in this approach has been proved to a large extent.

| | Total Weighted Knock-on Delay | | |
|---|---|---|---|
| | Initial Solution [-] | Optimized Solution [-] | Reduction [%] |
| Test Case 1 | 4905 | 3560 | -27.4 |
| Test Case 2 | 1936 | 1457 | -24.7 |
| Test Case 3 | 12468 | 3097 | -75.2 |
| Test Case 4 | 11194 | 2522 | -77.5 |
| Test Case 5 | 1858 | 637 | -65.7 |
| Test Case 6 | 2148 | 1277 | -40.6 |
| Test Case 7 | 2694 | 1894 | -29.7 |
| Test Case 8 | 1486 | 811 | -45.4 |

**Table 6-3: Experimental Results of the Test Case 1 – Test Case 8**

# 7  Summary and Further Research

In order to support dispatchers solving disturbances that occur during the operation process, a domain-specific dispatching optimization algorithm integrated in a multi-scale simulation model was developed in this dissertation. The whole workflow of the approach developed in this dissertation is illustrated in Figure 7-1.

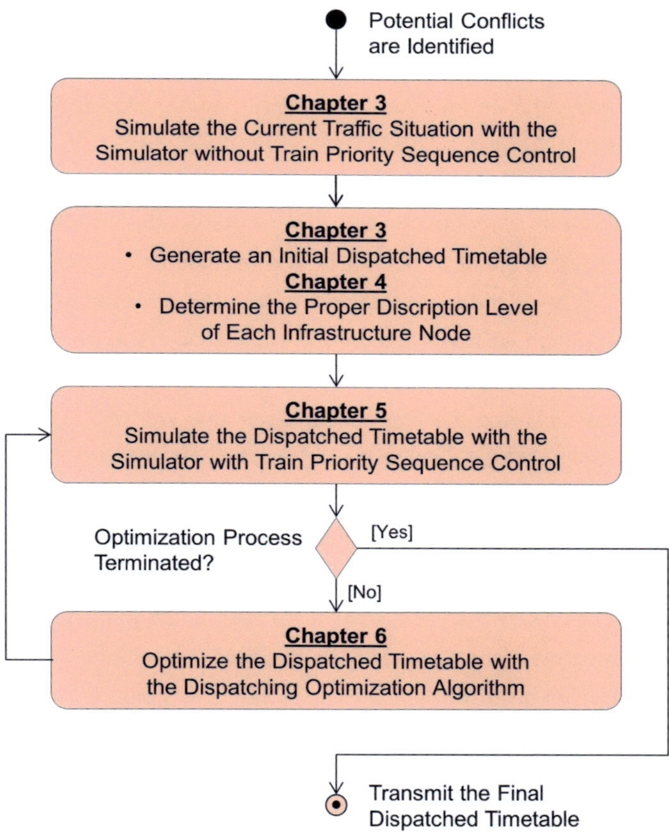


**Figure 7-1: The Complete Workflow of the Approach**

The main achievements are summarized as follows:

—  The multi-scale simulation model is characterized by continuous scaling, in which simulation is concurrently carried out on microscopic, mesoscopic and macroscopic levels (see Chapter 3). To enable the multi-scale simulation model to transform smoothly among different description levels, the developed microscopic, mesoscopic and macroscopic model followed the same framework of a synchronous approach. The concept of the multi-scale model contains two aspects: different levels of details of both the fundamental components constructing the simulation model and the interaction mechanism between the components. Furthermore, the simplest dispatching principle, First Come First Serve (FCFS), was implicitly implemented in the multi-scale model, and the model is capable of generating deadlock-free timetables independently, which are used as the basic timetable in further dispatching optimization processes.

– The appropriate description level of a specific area in the investigated area is determined by its significance value. Those areas with lower significance values are abstracted and simulated on the mesoscopic or macroscopic level; those with higher significance values are kept on the microscopic level. An assessment method was developed for the calculation of significant values (see Chapter 4). Two indicators are considered in the significant value: the relevance to conflicts and the aggregation accuracy. With the multi-scale model solution, computational complexity and accuracy of the simulation model are well-balanced.

– For the simulation of dispatched timetables and provision of necessary data for timetable characteristics evaluation in the optimization process, the multi-scale simulation was modified, and the function of train priority sequence control was integrated (see Chapter 5). In this model, train runs are regulated explicitly by a pre-given dispatched timetable. This model is a relatively independent module, which not only can be connected to the optimization model developed in this dissertation, but also can be integrated into other optimization models.

– The dispatching optimization model was developed based on tabu search algorithm, and the train priority sequences on loop tracks and open track sections adjusted through a series of dispatching actions in order to find an optimal dispatched timetable (see Chapter 6). To construct candidate solutions, an adaptive search strategy was adopted, in which both intensification and diversification elements were included in the candidate list, and their numbers dynamically adjusted according to the changes of solution quality. With this adaptive search strategy, the intensification and diversification searches are well-balanced.

- Knock-on delays were ranked according to their priorities to generate intensification elements (see Chapter 6). The priority of a knock-on delay is composed of two indicators: weighted knock-on delay and the influence of knock-on delay on further conflicts. To calculate the influence of knock-on delay on further conflicts, delay propagation was quantitatively modelled with the help of the simulation model. To resolve a selected knock-on delay, the suitable dispatching action was selected depending on its surrounding circumstances. Moreover, a train priority sequence adjustment method and a deadlock-free test were included in order to ensure that candidate solutions are deadlock-free.

The performance of the dispatching optimization algorithm was analyzed on a series of test cases with different traffic loads, and promising results were produced. In further researches, it would be meaningful to expand the multi-scale simulation model and the dispatching optimization model related to the following aspects:

- The multi-scale simulation model developed in this dissertation belongs to time-driven simulation. Event-driven simulation should be integrated to improve the calculation efficiency of the simulation model. Accordingly, the workflow of the simulation model should be modified to better integrate these two types of simulation mechanisms.
- The performance of the dispatching optimization algorithm can be further improved with the implementation of more reasonable dispatching rules and dynamic tabu list management. More reasonable dispatching rules could improve the quality of the initial solution, and the dynamic tabu tenure could increase the robustness of the tabu search algorithm.
- The potential disturbances, which may occur in the prediction time period, are not taken into account in the dispatching optimization process. The robustness of the dispatched timetable is questionable. Therefore, it is necessary to consider these potential disturbances in the dispatching process, and the efficiency and robustness of the dispatched timetable should be well balanced.

Railway dispatching is a comprehensive complex process. Besides the delay management discussed intensively in this dissertation, it also includes the related dispatching of train crews and rolling stocks. These specific constraints of railway operation should be modelled stepwise in simulation models. In future researches, not only the algorithm performance should be improved from the perspective of algorithm design, but also special attention should be paid to the latest technology development, in order to select out appropriate technologies and bring them into the field of railway operation and management.

# Abbreviations

| | |
|---|---|
| AA | Arrive-Arrive Headway |
| ATP | Automatic Train Protection |
| BS | Basic Structure |
| CG | Controlled Group |
| CTC | Centralized Traffic Control |
| DA | Depart-Arrive Headway |
| EG | Experimental Group |
| FCFS | First Come First Serve |
| FIFO | First In First Out |
| FRZ | Long Distance Passenger Train |
| GV | Freight Train |
| LT | Loop Track |
| LNT | Loop non Track |
| NRZ | Short Distance Passenger Train |
| OP | Operation Point |
| OT | Open Track Section |
| TS | Tabu Search |
| UIC | International Union of Railways |

## Symbols

| | |
|---|---|
| $a_{BS}$ | Amount of basic structures in the investigation area |
| $a_0, a_1, a_2, a_{2r}$ | Parameters of traction unit resistance |
| $a, b, c$ | Parameters of traction unit resistance |
| $a_{Br}$ | Braking acceleration rate [m/s$^2$] |
| $a_{Tr}$ | Acceleration rate [m/s$^2$] |
| $A_f$ | Cross-sectional area of the vehicles [m$^2$] |
| $AA_{Z_{Prev},Z_j,T}$ | Arrive-arrive headway between the train $Z_j$ and $Z_{Prev}$ on open track section $T$ |
| $ACC_{R_k,R_h}^{meso}$ | Aggregation accuracy of two occupation units $R_k$ and $R_h$ on the mesoscopic level |
| $ACC_N^{micro \rightarrow meso}$ | Aggregation accuracy of infrastructure node N from the microscopic level to the most detailed mesoscopic level |
| $c_a$ | Coefficient for axle adhesion |
| $c_b$ | Coefficient for the number of axles |
| $C_j$ | Constant for weighting knock-on delays of train $j$ |
| $C_j^t$ | Train-oriented indicator-based constant for train $j$ |
| $c_m$ | Value for air resistance |
| $C_{RED}$ | Value of possibility of reduction of delays |
| $C_{tw}$ | Relative importance of weighted knock-on delay |
| $DA_{Z_{Prev},Z_j,T}$ | Depart-arrive headway between the train $Z_j$ and $Z_{Prev}$ on open track section $T$ |
| $E(T_{t_2 \sim t_3,Z1,Z2}^{OVLP,R3})$ | Expected total overlapping time period between Z1 and Z2 |
| $\Delta E(T^{OVLP,R_k+R_h})$ | The relative change of expected total overlapping time period caused by combination of $R_k$ and $R_h$ |

| | |
|---|---|
| $F_{Rwp}$ | Vehicle resistance for passenger trains [N] |
| $F_R(v)$ | Train resistance at a given velocity $v$ [N] |
| $F_{Rt}(v)$ | Traction unit resistance at a given speed v [N] |
| $F_{Rlg}$ | Grade resistance of a train [N] |
| $F_{Tr}(v)$ | Tractive effort at wheel at a given velocity $v$ [N] |
| $g$ | Earth gravity constant 9.81 m/s$^2$ |
| $i[N]$ | Index of the macroscopic node N in the macro-path of a corresponding train |
| $Inf_{tw}$ | Influence of knock-on delay on further conflicts |
| $Inf_{tw_{j,i}}$ | Influence of a knock-on delay $tw_{j,i}$ on further conflicts |
| $Inf_{tw_{j,i}}'$ | Normalized influence of a knock-on delay $tw_{j,i}$ on further conflicts |
| $Inf_{min}$ | Minimum of the influences |
| $Inf_{max}$ | Maximum of the conflicts |
| $l_{Rn}$ | Length of the $n^{th}$ section of the train path |
| $L_{Rest}$ | Length of the rest of the train path |
| $m$ | Mass of the train [kg] |
| $m_T$ | Mass of traction unit [kg] |
| $m_w$ | Mass of all vehicles [kg] |
| $M$ | A number that is at least larger than the maximum of the aggregation accuracies |
| $n$ | Gradient [‰] or the length of the current block section list of the train |
| $n_{ges}$ | Total number of trains |
| $n_j^{block}$ | Number of block sections along the path of train $j$ |

| | |
|---|---|
| $n_w$ | Number of vehicles |
| $N^{Dispo}_{Z1,p}, N^{Dispo}_{Z2,q}$ | Index of the OT/LT in the dispatching relevant macro path of $Z_1$ or $Z_2$ |
| $N^{Dispo}_{Z1,p+1}, N^{Dispo}_{Z2,q+1}$ | Next dispatching relevant nodes determined for $Z_1$ and $Z_2$ |
| $OT_x$ | Open track section x |
| $\rho$ | Coefficient of increase in mass [-] |
| $\rho_T$ | Coefficient of increase in mass for a traction unit |
| $\rho_W$ | Coefficient of increase in mass for a vehicle |
| $P^{Rb}_{tw_{j,i},td_{l,k}}$ | Percentage of responsibility of delay $td_{l,k}$ for knock-on delay $tw_{j,i}$ |
| $P^{Rb}_{td_{j,k},tw_{j,i}}$ | Percentage of the responsibility of source knock-on delay $tw_{j,i}$ for delay $td_{j,k}$ |
| $P^{R3}_{t_2 \sim t_3, Zj}$ | Blocking probability of the time interval $[t_2, t_3]$ for $Zj$ on the occupation unit R3 |
| $P^{R3}_{t_2 \sim t_3, Z1,Z2}$ | Occurrence probability of the conflict situation between Z1 and Z2 on the occupation unit R3 in the time interval from $t_2$ to $t_3$ |
| $PI_{Zj,T/LT}$ | Original index of priority of a train $Z_j$ on T or LT |
| $Pri_{tw_{j,i}}$ | Priority of knock-on delay $tw_{j,i}$ |
| $S$ | Expected forward distance of the head of a train in the current time interval |
| $\Delta S$ | Distance between the current position of the train head and the brake application point |
| $S^{source}_{td_{j,k}}$ | Set of source knock-on delays of delay $td_{j,k}$ |
| $SV^{micro \rightarrow meso}_{N}$ | Significance value of an infrastructure node N to be abstracted from the microscopic level to the most detailed |

| | mesoscopic level |
|---|---|
| $SV_{R_k,R_h}^{meso}$ | Significance value of a large mesoscopic occupation unit composed of $R_k$ and $R_h$ on the mesoscopic level |
| $t_{t_2 \sim t_3,Z1,Z2}^{OVLP,R3}$ | Total overlapping time period between Z1 and Z2 |
| $td_{j,k}$ | Delay of train $j$ on block section $k$ |
| $t_{h_{j,i,l,k}}$ | Time period during which train $j$ on block $i$ had been hindered by train $l$ on block $k$ |
| $t_{h_{j,i}}$ | Time period during which train $j$ on block $i$ had been hindered by the other trains |
| $t_{j,i}^{start,Ist}$ | Actual start blocking time of train $j$ on occupation unit $i$ |
| $t_{j,i}^{start,Soll}$ | Scheduled start blocking time of train $j$ on occupation unit $i$ |
| $tw_{j,i}$ | Knock-on delay of train $j$ on block section $i$ |
| $tw_{j,i}^{CG}$, $tw_{j,i}^{EG}$ | Knock-on delay of train $j$ on block section $i$ in the controlled group (CG) or experimental group (EG) |
| $\widehat{tw}_{j,i}$ | Weighted knock-on delay of train j on block section i |
| $\widehat{tw}_{j,i}'$ | Normalized weighted knock-on delay of train $j$ on block section $i$ |
| $\widehat{tw}_{min}$ | Minimum of the weighted knock-on delays |
| $\widehat{tw}_{max}$ | Maximum of the weighted knock-on delays |
| $T_{now}$ | Current execution time in the simulation model |
| $t_{Pein\,j}$ | Original delay of train $j$ |
| $t_{Pur\,j}$ | Initial delay of train $j$ |
| $t_{R\,j,i}$ | Reserve time of train $j$ on block section $i$ |
| $TB_{i[N]-1,Z_j}$ | Departing/passing time for train $Z_j$ in the $(i[N]-1)^{th}$ node of its macro-path |

| | |
|---|---|
| $TI_{i[N],Z_j}$ | Scheduled operation time for train $Z_j$ in the $(i[N])^{th}$ node of its macro-path |
| $v_r$ | Relative speed between air and the train [km/h] |
| $v_{max}$ | Maximum allowed speed for a train at a certain time instant |
| $v_{max,Train}$ | Maximum speed of the train |
| $v_{max,Block\ i}$ | Maximum speed of the current block section $i$ of the train |
| $zj$ | Amount of block sections along the path of train $j$ |
| $Zfl$ | Viscosity of a certain dispatching condition |
| $Z_j$ | Train $Z_j$ |
| $Z_{Prev}$ | Immediately previous train |

# Appendix I: Estimation of Forward Distance on Microscopic Level

The forward distances of train heads need to be estimated both for the detection of new resource requirement and the update of train positions in a time interval. The algorithm for estimating forward distance in one time interval on the microscopic level, which is to be described in this section, was developed within the scope of the DFG project [Martin and Liang, 2017].

The movement behavior of a train is restricted by the attributes of infrastructure and the train itself, along with operational constraints (e.g. brake application points and stopping points). Three behavior sections are considered in this dissertation: the acceleration section, constant movement section and braking section. Furthermore, two basic methods are developed to assist the estimation of forward distance: the forward distance estimated based on the acceleration section and/or constant movement (abbreviation: ForwardDist_AccConst), and the forward distance estimated based on the braking section (abbreviation: ForwardDist_Brake).

The method ForwardDist_AccConst is designed to calculate the forward distance of a train in one time interval, supposing that the train will accelerate in the current time interval and turn into constant movement if the maximum allowed speed is reached. Special attention must be given to the fact that the ForwardDist_AccConst method only provides intermediate results, which are used to roughly estimate the train head position at the end of each time interval. The results may have to be fine-tuned with consideration of the other constraints. In the ForwardDist_AccConst method, the maximum allowed speed should be determined primarily, which is the minimum value between the maximum speed of the train and the maximum speed limit of each current block section.

$$v_{max} = \min \left\{ v_{max,Train}, \; v_{max,Block\,1}, \cdots, v_{max,Block\,i}, \cdots, v_{max,Block\,n} \right\} \qquad \text{(I-1)}$$

Notation used:

$v_{max}$:　　　　　　　maximum allowed speed for a train at a certain time instant

$v_{max,Train}$:　　　　　maximum speed of the train

$v_{max,Block\,i}$:　　　　maximum speed of the current block section $i$ of the train

$n$:                    the length of the current block section list of the train

The acceleration rate $a_{Tr}$ can be calculated based on the speed of the train at the beginning of the current time interval $v_0$ with Formula (3-8), and the speed of the train at the end of the current time interval $v_t$ can be calculated with Newton formula:

$$v_t = v_0 + a_{Tr} \cdot t \cdot 3.6 \qquad \text{(I-2)}$$

The length of a time interval $t$ is taken as one second, and in such a short time the acceleration rate can be assumed constant. The speed difference should be transformed into km/h (1 m/s = 3.6 km/h). If $v_t$ is equal or smaller than $v_{max}$, it implies that the train can keep accelerating in the current time interval. The speed $v_t$ is the estimated speed at the end of the current time interval, and the forward distance in the current time interval $S$ can be calculated as:

$$S = \frac{v_t^2 - v_0^2}{2 \cdot a_{Tr} \cdot 3.6} \qquad \text{(I-3)}$$

Furthermore, the forward distance is composed of two sections, as shown in Appendix Figure 1. The train accelerates until $v_{max}$ is reached in the first section, and then runs at a constant speed of $v_{max}$ in the second section. The forward distance in the first section $S_1$ can be calculated with Formula (I-3), and only $v_t$ needs to be replaced by $v_{max}$.

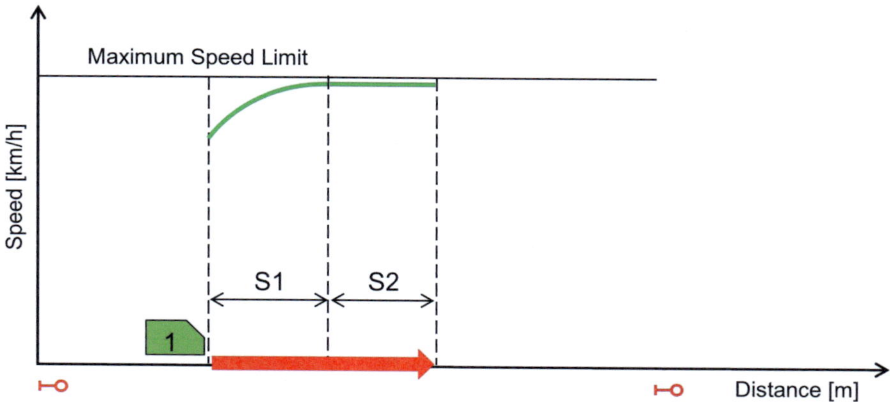


**Appendix Figure 1: Forward Distance Estimation – Acceleration Section and Constant Movement**

The first section takes $t_1$ second, and the second section takes $t_2$ second.

$$t_1 = \frac{v_{max} - v_0}{a_{Tr} \cdot 3.6}$$

(I-4)

$$t_2 = t - t_1$$

(I-5)

So the forward distance in the second section is

$$S_2 = v_{max} \cdot t_2$$

(I-6)

Finally, the forward distance in the current time interval $s$ can be determined.

$$S = S_1 + S_2$$

(I-7)

In case of pure constant movement, because $v_0$ is equal to $v_{max}$, $t_1$ and $S_1$ are equal to zero. The forward distance $S$ is equal to $S_2$. So the ForwardDist_AccConst method is a general approach to calculate the forward distance in case of acceleration section AND-OR constant movement.

The method ForwardDist_Brake is used to estimate the forward distance of a train in one time interval when the train only brakes. At first the braking acceleration rate $a_{Br}$ should be calculated with Formula (3-10). The speed of the train at the end of the time interval $v_t$ can be calculated:

$$v_t = v_0 + a_{Br} \cdot t \cdot 3.6 \tag{I-8}$$

If $v_t$ is equal to or greater than zero, it can be used as the train speed at the end of the current time interval; otherwise, $v_t$ should be set to zero (Appendix Figure 2). In both cases the forward distance can be calculated with Formula (I-9).

$$S = \frac{v_t^2 - v_0^2}{2 \cdot a_{Br} \cdot 3.6} \tag{I-9}$$

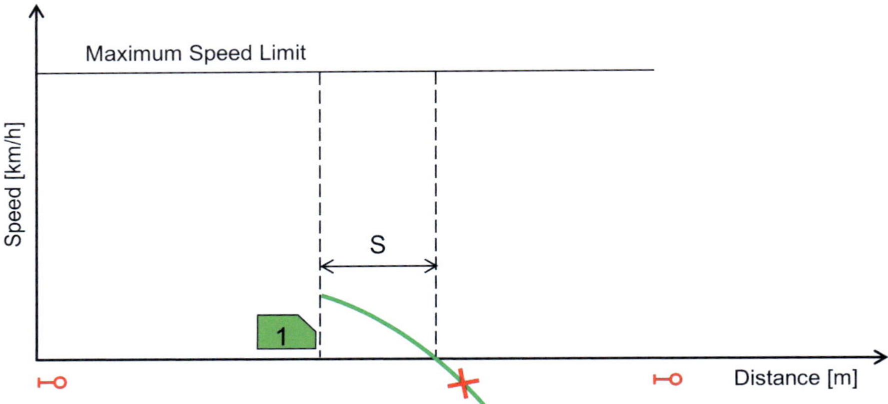


**Appendix Figure 2: Forward Distance Estimation – Braking Section**

The forward distance estimated with the basic methods described above may have to be fine-tuned when additional constraints exist (e.g. speed reduction on further path and scheduled or unscheduled stops). Due to the different speed control mechanisms of intermittent and continuous ATP systems, their fine-tuning procedures will be elaborated separately in the following context.

For a train under the mode of intermittent ATP system, if the current speed of the train is equal to zero, the forward distance and speed estimated with the method ForwardDist_AccConst can be directly considered as the final estimation results and further fine-tuning is not necessary. If the current speed is not equal to zero, the procedures to be carried out are dependent on the current position of the train head, of which exist two possibilities: before or beyond the distant signal for the next block section.

The distance between the distant signal for the next block section and the rear signal of the last current block section is great enough that the train cannot run through it in one time interval. Therefore, in case the current position of the train head is before the distant signal, only the constraints (i.e. maximum speed limit and brake application point) on the current block sections should be considered. Furthermore, in this case the train head has not yet absolutely passed the brake application point located in the last current block section, since brake application points theoretically ought to be located between the distant signal for the next block section and the rear signal of the current block section. If the position of the train head will not exceed the brake application point[27] located in the last current block section, the forward distance estimated with the ForwardDist_AccConst method can be directly considered as the final result; otherwise the braking section should be taken into account, and the result should be correspondingly adjusted.

In order to estimate the forward distance accurately, the movement behavior of the involved train in the current time interval should analyzed. Beyond the brake application point only the braking section is involved, while before the brake application point the movement behavior can be composed of the constant movement (Appendix Figure 3) or acceleration sections (Appendix Figure 4), or a combination of them (Appendix Figure 5). These three cases will be explained separately due to their different calculation processes.

In order to determine concretely that a situation belongs to a given case, firstly, the acceleration rate of the train $a_{Tr}$ should be calculated with Formula (3-8). If $a_{Tr}$ is equal to zero, it implies that the train runs at a constant speed before the application point (Appendix Figure 3). The distance between the current position of the train head and the brake application point is designated as $\Delta S$. The time duration of the constant movement $t_1$ is $\Delta S/v_0$, and the time duration of braking section $t_2$ is $(t - t_1)$. The train speed at the end of the first section $v_1$ is equal to $v_0$. The braking

---

[27] The brake application point in a certain block section is expressed as the distance from the entrance signal of the block section to the brake application point. The default value of the distance is positive infinite. So, if no concrete brake application point exists in a certain block section, the involved train will never exceed the brake application point.

acceleration rate $a_{Br}$ based on $v_1$ can be calculated with Formula (3-10). The adjusted train speed at the end of the current time interval and the forward distance in the braking section are

$$v_t = v_1 + a_{Br} \cdot t_2 \cdot 3.6 \tag{I-10}$$

$$S_2 = \frac{v_t^2 - v_1^2}{2 \cdot a_{Br} \cdot 3.6} \tag{I-11}$$

and the adjusted forward distance of the train head in the current time interval is

$$S = S_1 + S_2 = \Delta S + S_2 \tag{I-12}$$

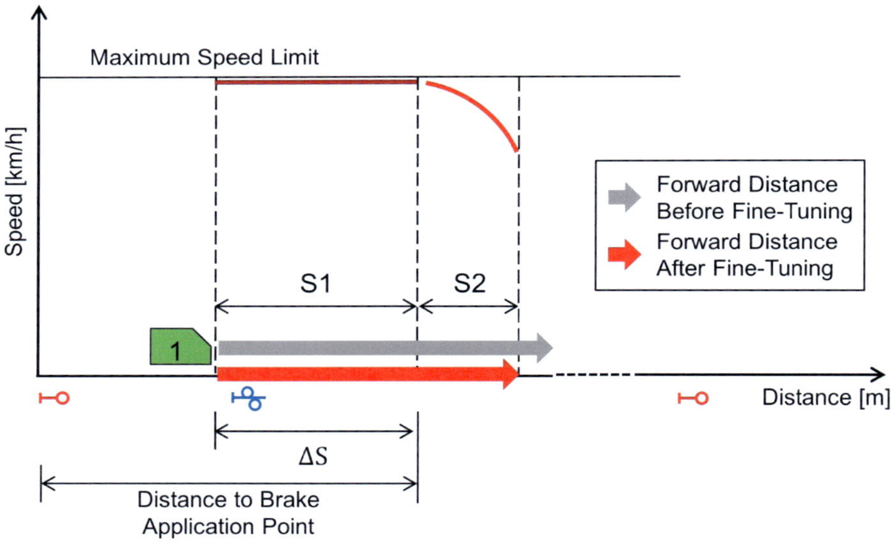


Appendix Figure 3: Forward Distance Fine – Tuning – Intermittent ATP System (I)

If the acceleration rate $a_{Tr}$ calculated based on $v_0$ is not equal to zero, the expected speed of the train at the brake application point $v_1$ can be derived from Formula (I-3):

$$v_1 = \sqrt[2]{2 \cdot a_{Tr} \cdot \Delta S + v_0^2} \tag{I-13}$$

If $v_1$ is smaller than or equal to $v_{max}$ (see method ForwardDist_AccConst), the movement behavior of the train includes two sections: the acceleration section and the braking section. The time duration of each section is calculated as:

$$t_1 = \frac{v_1 - v_0}{2 \cdot a_{Tr} \cdot 3.6} \tag{I-14}$$

$$t_2 = t - t_1 \tag{I-15}$$

The train speed at the end of the current time interval and the forward distance in the braking section can be calculated with Formulas (I-10) and (I-11), and the adjusted forward distance in the current time interval can be determined with Formula (I-12).

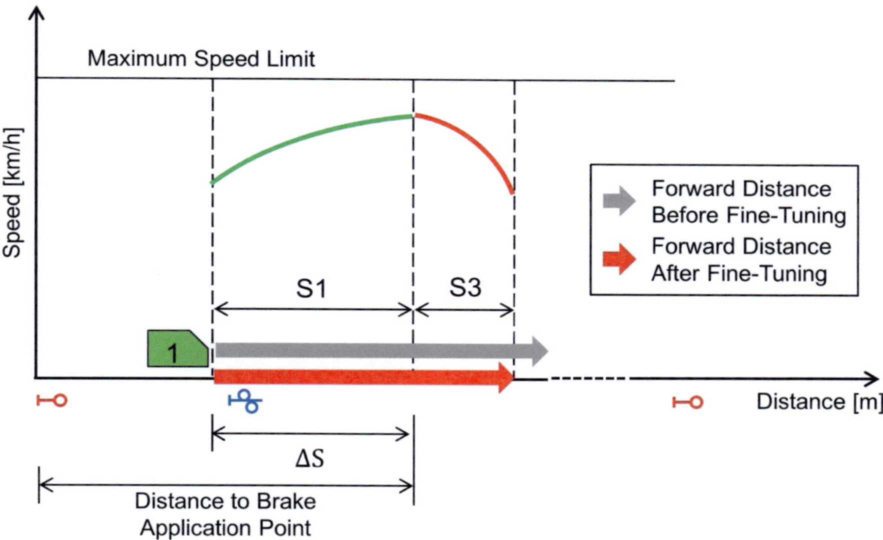


**Appendix Figure 4: Forward Distance Fine – Tuning – Intermittent ATP System (II)**

If $v_1$ is larger than $v_{max}$, it implies that the train speed will reach $v_{max}$ before the brake application point. Therefore, the movement behavior in the current time interval should include three sections: the acceleration section, the constant movement section and the braking section (Appendix Figure 5). The time duration of the acceleration section $t_1$ and the corresponding forward distance $S_1$ are

$$t_1 = \frac{v_{max} - v_0}{a_{Tr} \cdot 3.6} \tag{I-16}$$

$$S_1 = \frac{v_{max} - v_0}{2 \cdot a_{Tr} \cdot 3.6} \cdot t_1 \tag{I-17}$$

The forward distance in the section of constant movement $S_2$ is $(\Delta S - S_1)$, and the time duration $t_2$ is $(\Delta S - S_1)/v_{max}$. The time left for the braking section $t_3$ is $(t - t_1 - t_2)$. The train speed at the end of the current time interval $v_t$ and the forward distance in the braking section $S_3$ are

$$v_t = v_{max} + a_{Br} \cdot t_3 \cdot 3.6 \tag{I-18}$$

$$S_3 = \frac{v_t^2 - v_{max}^2}{2 \cdot a_{Br} \cdot 3.6} \tag{I-19}$$

The adjusted forward distance in the current time interval can be obtained:

$$S = S_1 + S_2 + S_3 = \Delta S + S_3 \tag{I-20}$$

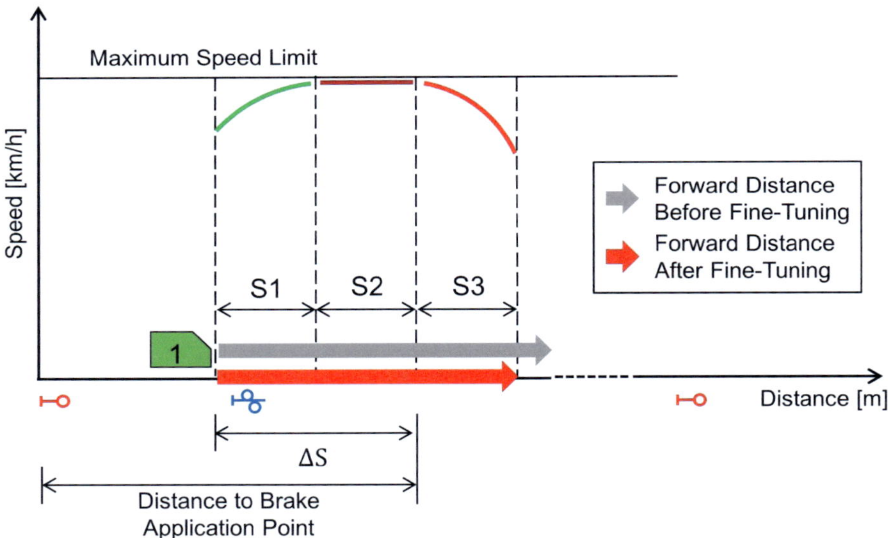


**Appendix Figure 5: Forward Distance Fine – Tuning – Intermittent ATP System (III)**

Under the condition that a train has to start braking from a certain position, there are three possible consequential situations as described above. No matter where the train is located and which ATP system the train is using, in principle the procedures of fine-tuning illustrated in Appendix Figure 3, Appendix Figure 4 and Appendix Figure 5 are generally applicable, and only minor details should be adjusted. Therefore, these procedures will be designated as standard fine-tuning procedures for brake application in the following text.

In case the current position of the train head is beyond the distant signal for the next block section, it is possible that the current position of the train head has already exceeded the brake application point in the last current block section. If so, the forward distance in the current time interval and the train speed at the end of the current time interval can be determined with the ForwardDist_Brake method, and the results can be directly considered as the final estimation results. If the train head has not yet passed the brake application point, its expected position should be roughly calculated with the ForwardDist_AccConst method as usual.

If the expected position of the train head exceeds the brake application point, the forward distance and the corresponding speed can be fine-tuned with the fine-tuning procedures for brake application; otherwise, the results calculated by the ForwardDist_AccConst method can be directly considered as the final estimation results, in case the expected position of train head does not leave the last current block section.

Upon the condition that the expected position of train head enters the next block section but does not pass the brake application point in the last current block section (no concrete brake application point exists), the roughly estimated forward distance may have to be recalculated, because the speed constraints on the next block section are involved. Only if the maximum speed limit of the next block section is reduced and the expected speed of the train is over the speed limit (Appendix Figure 6), should the forward distance be recalculated.

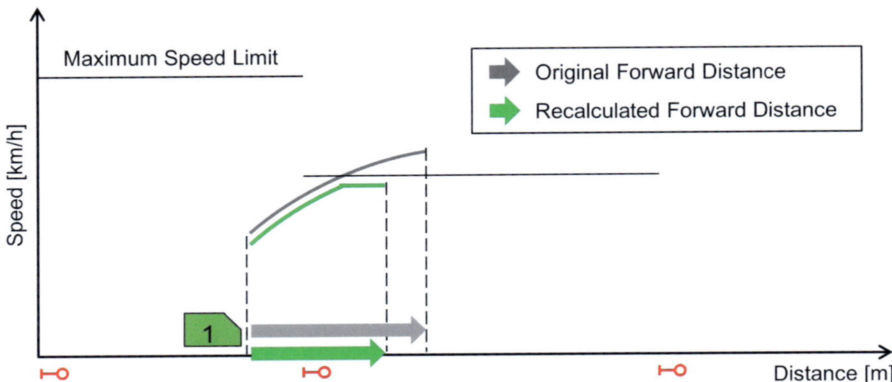


**Appendix Figure 6: Recalculation of Forward Distance**

Because the train is going to enter the new block section, it should be checked that whether the train head passes the brake application point in the next block section. If so, the forward distance and speed should be further adjusted with the standard fine-tuning procedures for brake application; otherwise the roughly estimated results can be directly considered.

For a train under the mode of continuous ATP system, if the current speed of the train is equal to zero, the results estimated with the ForwardDist_AccConst method can be directly considered; otherwise, it is necessary to detect the existence of stopping points within the nominal braking distance of the train, and the estimated forward distance may have to be fine-tuned accordingly.

Similar to resource requirement in continuous ATP territory, the nominal stopping point of a train is also used herein to detect stopping points (i.e. stopping points for speed reduction and temporal stopping points). The positions of the nominal stopping point at the beginning and at the end of the current time interval should be determined primarily as shown in Appendix Figure 7. It is assumed that the nominal stopping point is shifted from Block A to Block B in the current time interval (Block A and Block B can be the same block section). There is likely more than one stopping point located in Block A and Block B, but only the stopping point nearest to the current position of train head needs to be taken into account. On one hand, the nearest stopping point is the most urgent. On the other hand, the nominal stopping point has no chance to exceed the nearest stopping point before it is removed, so the further stopping points cannot influence the movement behavior of the train in the current time interval.

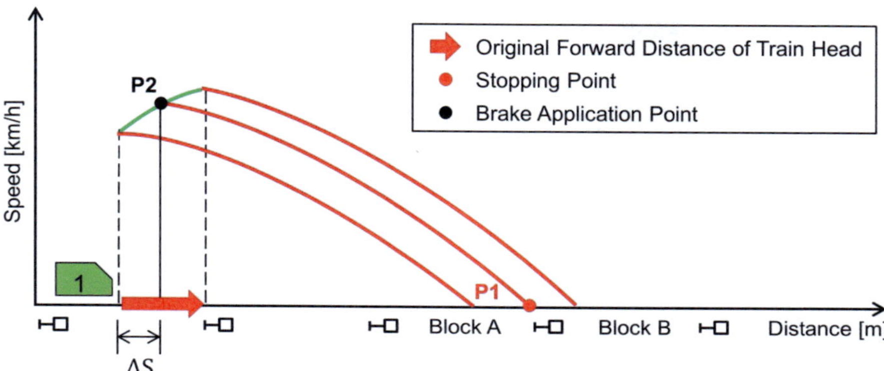


**Appendix Figure 7: Forward Distance Fine-Tuning – Continuous ATP System**

In the condition that the nominal stopping point passes the nearest stopping point (P1) in the current time interval, it is implied that the train must start braking from a certain point. The optimal brake application point is the intersecting point (P2) between the originally expected speed-distance curve for the acceleration section and/or the constant movement (green curve) and the braking curve deduced from the nearest stopping point (red curve between P1 and P2). The distance between the current position of the train head and the optimal brake application point can be easily determined (ΔS), with which the forward distance in the current time interval and the speed of the train at the end of the current time interval can be adjusted, following the standard fine-tuning procedures for brake application. In case the nominal stopping point does not pass any stopping point in the current time interval, the results estimated with the ForwardDist_AccConst method can be directly considered.

# Appendix II: Performance of Dispatching Optimization Algorithm on Test Cases

### Test Case 3: 22 Trains/Hour

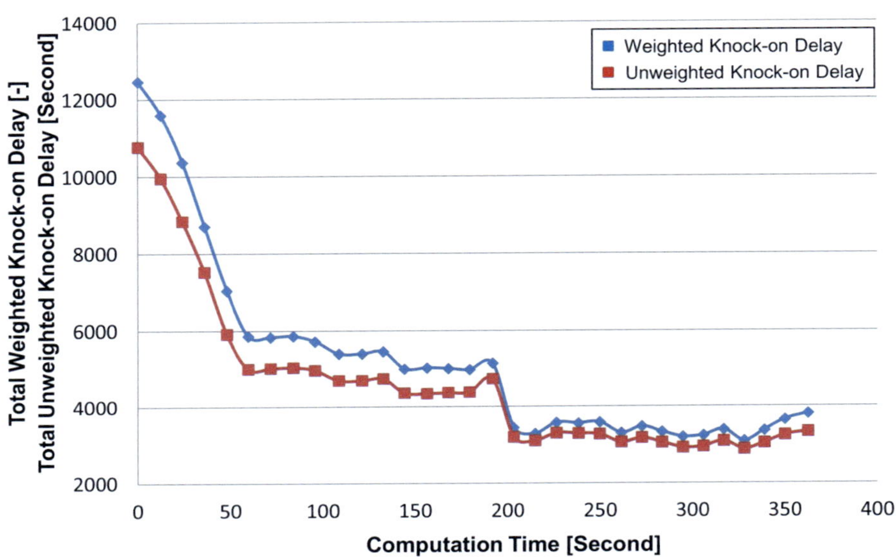


### Test Case 4: 20 Trains/Hour

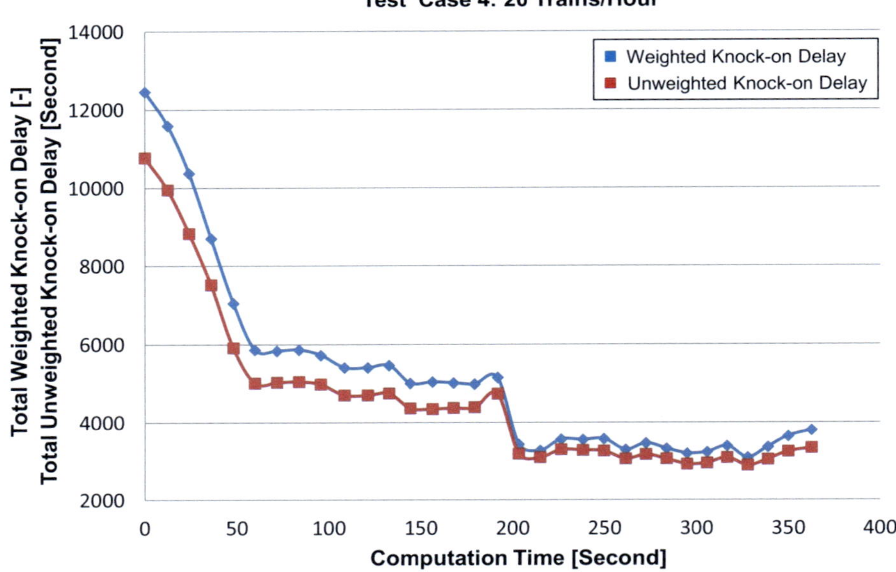

### Test Case 5: 19.5 Trains/Hour

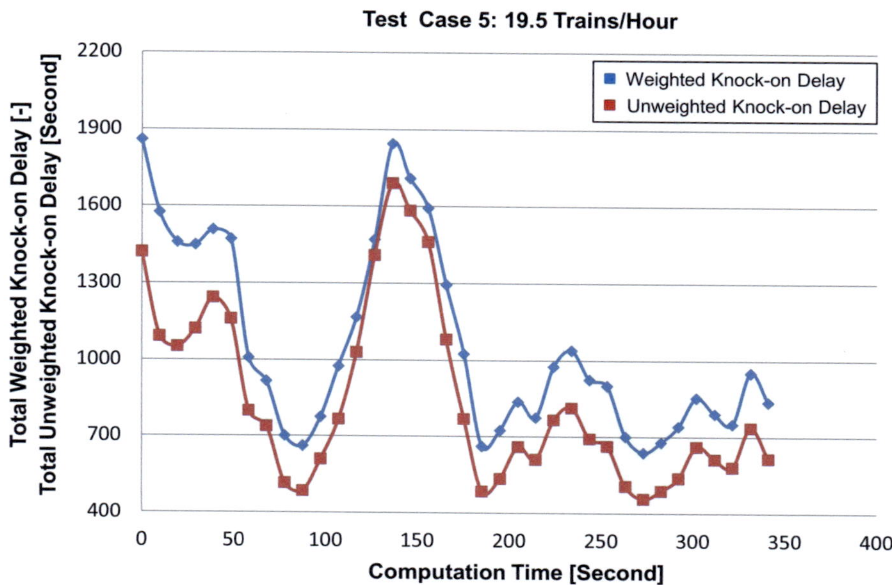



### Test Case 6: 17 Trains/Hour

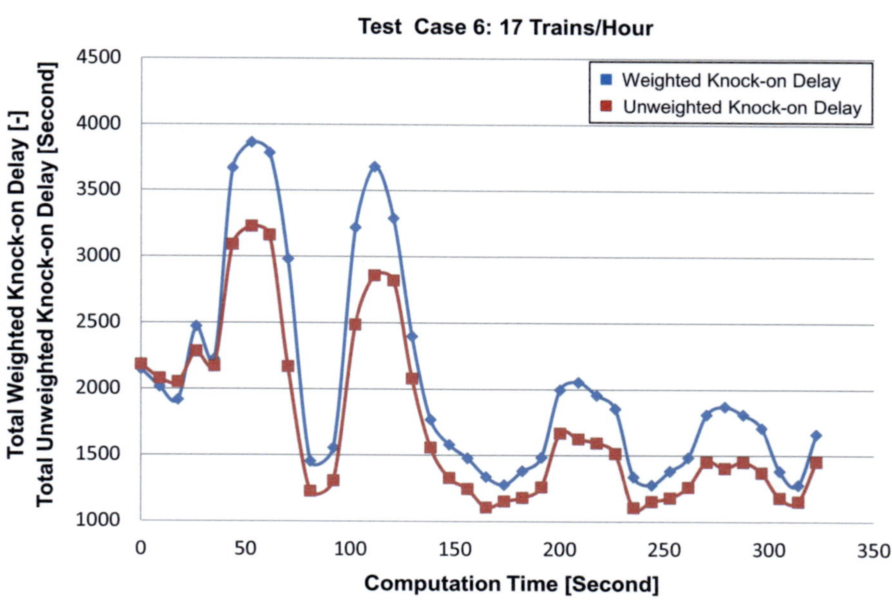

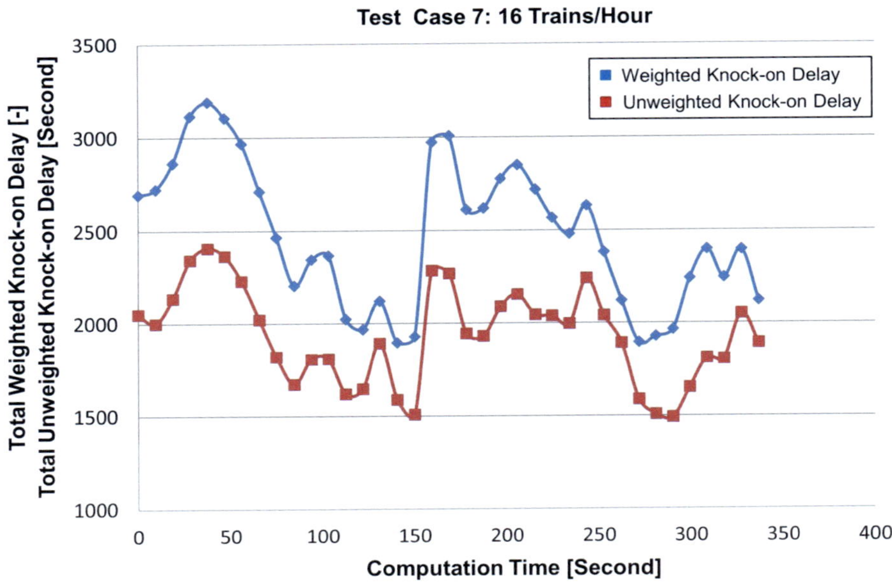



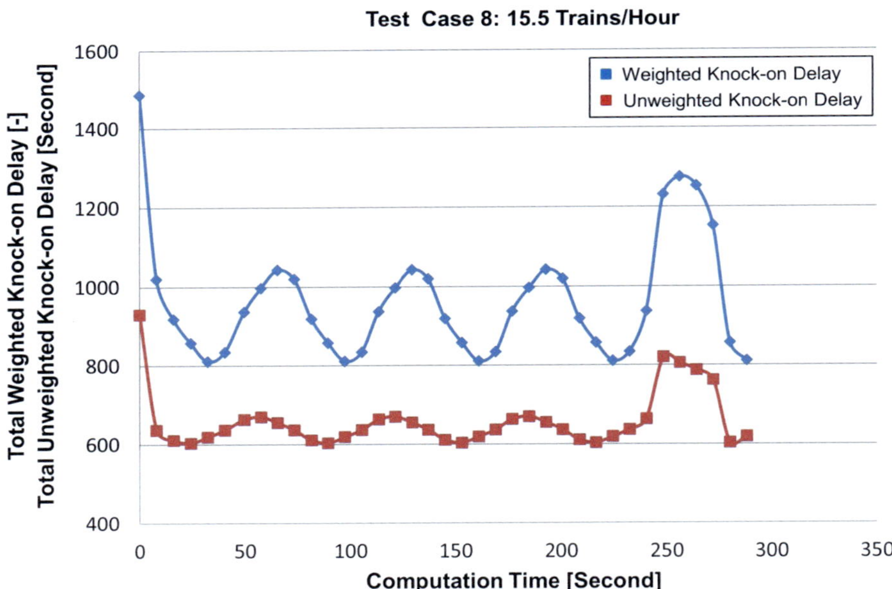



**Appendix Figure 8: Performance of Dispatching Optimization Algorithm on Test Case 3-8**

# Bibliography

[Alwadood et al., 2012]

Alwadood, Zuraida; Shuib, Adibah; Abd.Hamid, Norlida: *A Review on Quantitative Models in Railway Rescheduling*. In: International Journal of Science & Engineering Research, Vol. 3, Issue 6, 2012

[Brünger and Dahlhaus, 2014]

Brünger, Olaf; Dahlhaus, Elias: *Running Time Estimation*. In: Ingo Arne Hansen, Jörn Pachl (eds.): Railway Timetabling & Operations, Eurailpress, Hamburg, 2014

[Cacchiani et al., 2013]

Cacchiani, Valentina; Huisman, Dennis; Kidd, Martin; Kroon, Leo; Toth, Paolo; Veelenturf, Lucas; Wagenaar, Joris: *An Overview of Recovery Models and Algorithms for Real-time Railway Rescheduling*. In: Econometric Institute Report EI2013-29, 2013

[Chu, 2014]

Chu, Zifu: *Modellierung der Wartezeitfunktion bei Leistungsuntersuchungen im Schienenverkehr unter Berücksichtigung der transienten Phase*. Dissertation, Universität Stuttgart, 2014

[Corman and Meng, 2013]

Corman, Francesco; Meng, Lingyun: *A Review of Online Dynamic Models and Algorithms for Railway Traffic Control*. In: IEEE, 2013.

[Corman et al., 2011a]

Corman, F.; D'Ariano, A.; Pacciarelli, D.; Pranzo, M.: *Dispatching and Coordination in Multi-area Railway Traffic Management*. RT-DIA-190, 2011

[Corman et al., 2011b]

Corman, F.; D'Ariano, A.; Pranzo, M.; Hansen, I.A.: *Effectiveness of Dynamic Reordering and Rerouting of Trains in a Complicated and Densely Occupied Station Area*. In: Transport planning and technology, Vol.34, Issue 4, 2011, pp. 341-362

[Corman et al., 2010]

Corman, F.; D'Ariano, A.; Pacciarelli, D.; Pranzo, M.: *Centralized versus Distributed Systems to Reschedule Trains in two Dispatching Areas.* Public Transport, Vol.2, Issue 3, 2010, pp. 219-247

[Corman et al., 2009]

Corman, F.; Goverde, R.M.P.; and D'Arinao, A.: *Rescheduling Dense Train Traffic over Complex Station Interlocking Areas.* In: Springer-Verlag Berlin Heidelberg, 2009

[Cui et al., 2012]

Cui, Yong; Krohn, Teresa; Martin, Ullrich; Tritschler, Stefan: *Consistent Decision Process and Algorithm for Train Dispatching.* In: Network for Mobility, 6th International Symposium, Stuttgart, Germany, 2012

[Cui and Martin. 2011]

Cui, Yong; Martin, Ullrich: *Multi-scale Simulation in Railway Planning and Operation.* In: Promet-Traffic and Transportation, Vol.23, No.6, 2011, pp. 511-517

[Cui, 2010]

Cui, Yong: *Simulation-Based Hybrid Model for a Partially Automatic Dispatching of Railway Operation.* Dissertation, Universität Stuttgart, 2010

[Cui, 2005]

Cui, Yong: *Implementation of the Optimization Theory for User Oriented Automatic Dispatching Systems in Railway Transport.* Master Thesis, Universität Stuttgart, 2005

[DB NETZ AG, 420.0105]

DB NETZ AG: Richtlinie 420.0105 – Dispositionsregeln, 2005

[DB NETZ AG, 420.02]

DB NETZ AG: Richtlinie 420.02, 2010

[D'Ariano, 2008]

D'Ariano, A.: *Improving Real-Time Train Dispatching: Models, Algorithms and Applications.* PhD Thesis, TU Delft, 2008

[D'Ariano and Pranzo, 2008]

D'Ariano, A.; Pranzo, M.: *An Advanced Real-time Train Dispatching System for Minimizing the Propagation of Delays in A Dispatching Area under Severe Disturbances.* In: Networks and Spatial Economics, Vol.9, Issue 1, 2008, pp. 63-84

[Fabris et al., 2014]

Fabris, Stefano; Longo, Giovanni; Medeossi, Giorgio; Pesenti, Raffaele: *Automatic Generation of Railway Timetables based on A Mesoscopic Infrastructure Model.* In: Journal of Railway Transport Planning & Management, Vol. 4, Issues 1-2, August-October 2014, pp. 2-13

[Fan, 2012]

Fan, Bo: *Railway Traffic Rescheduling Approaches to Minimize Delays in Disturbed Conditions.* Dissertation, University of Birmingham, UK, 2012

[Fay, 2000]

Fay, Alexander: *A Fuzzy Knowledge-based System for Railway Traffic Control.* In: Engineering Applications of Artificial Intelligence, Vol. 13, Issue 6, December 2000, pp. 719-729

[Glover et al., 2007]

Glover, Fred; Laguna, Manuel; Marti, Rafael: *Principles of Tabu Search.* In: Handbook on Approximation Algorithms and Metaheuristics, 2007

[Glover and Laguna, 1993]

Glover, Fred; Laguna, Manuel: *Tabu Search.* In: Modern Heuristic Techniques for Combinatorial Problems, 1993

[Glover, 1986]

Glover, F.: *Future Paths for Integer Programming and Links to Artificial Intelligence.* In: Computer and Operations Research, Vol. 13, No. 5, 1986, pp. 533-549

[Goverde, 2010]

Goverde, R.M.P.: *A Delay Propagation Algorithm for Large-Scale Railway Traffic Networks.* In: Transportation Research Part C, Vol.18, Issue 3, 2010, pp. 269-287

[IEV, 2011]

IEV: *Project "Kosidispo" Konsistente Disposition in Planung und Betrieb AP1: Algorithmierung Dispositionsregeln.* Universität Stuttgart, 2011

[Jacobs, 2008]

Jacobs, Jürgen: *Rescheduling.* In: Ingo Arne Hansen, Jörn Pachl (eds.): Railway Timetabling and Traffic, Eurailpress, Hamburg, 2008

[Kecman, 2014]

Kecman, Pavle: *Models for Predictive Railway Traffic Management.* PhD Thesis, TU Delft, 2014

[Kecman et al., 2012]

Kecman, Pavle; Corman, Francesco; D'Ariano, Andrea, Goverde, Rob M.P.: *Rescheduling Models for Network-wide Railway Traffic Management,* Research Project: Model-Predictive Railway Traffic Management (Project No. 11025), 2012

[Kettner et al., 2003]

Kettner, Michael; Sewcyk, Bernd; Eickmann, Carla: *Integrating Microscopic and Macroscopic Models for Railway Network Evaluation.* In: Proceedings of the European Transport Conference (ETC), Strasbourg, France, 2003

[Larsen et al., 2013]

Larsen, R.; Pranzo, M.; D'Ariano, A.; Pacciarelli, D.; Corman, F.: *Susceptibility of Optimal Train Schedules to Stochastic Disturbances of Process Times.* In: Journal of Flexible Services and Manufacturing, 2013.

[Liu et al., 2014]

Liu, Guangyuan; He, Yi; Wen, Wanhui: *Tabu Search Algorithm and Its Application.* Science Press, 2014

[Luethi et al., 2007]

Lüthi, M.; Nash, A.; Weidmann, U.; Laube, F.; Wüst, R.: *Increasing Railway Capacity and Reliability through Integrated Real-Time Rescheduling.* In: Proceedings of the 11th World Conference on Transport Research, Berkeley, 2007

[Marinov and Viegas, 2011]

Marinov, Marin; Viegas, Jose: *A Mesoscopic Simulation Modelling Methodology for Analyzing and Evaluating Freight Train Operations in Railway Network*. In: Simulation Modelling Practice and Theory, 2011

[Martin and Liang, 2017]

Martin, Ullrich; Liang, Jiajian: *The Influence of Dispatching on the Relationship between Capacity and Operation Quality of Railway Systems*. Neues verkehrswissenschaftliches Journal – NVJ, Verkehrswissenschaftliches Institut Stuttgart, 2017 (in preparation)

[Marin et al., 2015]

Martin, Ullrich; Liang, Jiajian; Cui, Yong: *Einfluss von ausgewählten Dispositionsparametern auf das Ergebnis von Leistungsuntersuchungen*. In: ETR-Eisenbahntechnische Rundschau, Nr. 7+8, Jr. 64, 2015, pp. 20-23.

[Martin, 2014]

Martin, Ullrich: *Performance Evaluation*. In: Ingo Arne Hansen, Jörn Pachl (eds.): Railway Timetabling and Traffic, Eurailpress, Hamburg, 2014

[Martin et al., 2012]

Martin, Ullrich; Li, Xiaojun; Warninghoff, Carsten-Rainer: *Bewertungsverfahren für Knotenelemente bei der Infrastrukturbemessung-RePlan*. In: ETR-Eisenbahntechnische Rundschau, Nr. 11, 2012, pp. 38-43.

[Martin et al., 2008a]

Martin, Ullrich; Li, Xiaojun; Schmidt, Christine: *PULEIV Projektbericht. Allgemeingültiges Verfahren zur praxisorientierten Bestimmung des Leistungsverhaltens von Eisenbahninfrastrukturen (unveröffentlicht)*. Im Auftrag der DB Netz AG., 2008

[Martin et al., 2008c]

Martin, Ullrich; Li, Xiaojun; Schmidt, Christine et al.: *PULEIV Benutzerhandbuch*. Version 1.0.7 (unveröffentlicht), 2008

[Martin et al., 2008d]

Martin, Ullrich; Li, Xiaojun; Schmidt, Christine et al.: *PULEIV Anwenderleitfaden*. Version 1.0.7 (unveröffentlicht), 2008

[Martin, 2002]

Martin, Ullrich: *Deviation Management in Rail Operation*. In: Networks for Mobility, Proceedings of the International Symposium, 2002

[Martin, 1995]

Martin, Ullrich: *Verfahren zur Bewertung von Zug- und Rangierfahrten bei der Disposition*. Dissertation. TU Braunschweig, 1995

[Molo, 2017]

Von Molo, Carlo: *Projekt "Regler" – Zusammenfassender Rückblick auf die Testphase in der BZ Karlsruhe Januar 2012 bis März 2013*, Project Final Report, 2017 (to be published).

[OpenTrack, 2016]

OpenTrack: *OpenTrack – Simulation of Railway Networks, 2016*.

[Pachl, 2016]

Pachl, Jörn: *Glossary of Railway Operation and Control* (03.11.2016). <http://www.joernpachl.de/glossary.htm>, Accessed 03.November 2016.

[Pachl, 2014]

Pachl, Jörn: *Timetable Design Principles*. In: Ingo Arne Hansen, Jörn Pachl (eds.): Railway Timetabling & Operations, Eurailpress, Hamburg, 2014

[Pachl, 2011]

Pachl, Jörn: *Deadlock Avoidance in Railroad Operations Simulations*. In: 90th Annual Meeting of Transportation Research Board in Washington DC, 2011

[Pachl, 2002]

Pachl, Jörn: *Railway Operation and Control*. VTD Rail Publishing, 2002

[Pachl, 1993]

Pachl, Jörn: *Steuerlogik für Zuglenkanlagen zum Einsatz unter stochastischen Betriebsbedingungen.* TU Braunschweig, 1993

[Radtke, 2014]

Radte, A.: *Infrastructure Modeling,* In: Ingo Arne Hansen, Jörn Pachl (eds.): Railway Timetabling & Operations, Eurailpress, Hamburg, 2014

[PT1, 2016]

Public Transportation and Railway Operation. Universität Stuttgart, Lecture Note, Winter Semester, 2016

[Quaglietta, 2011]

Quaglietta, Egidio: *A Microscopic Simulation Model for Supporting The Design of Railway Systems: Development and Application.* PhD Thesis, University of Naples Federico II, 2011

[RMCon, 2016]

RMCon: *Railsys Suite – Innovation IT Solution for Railway Transport,* 2016

[RMCon, 2007]

RMCon Software: Railsys-7.6.12, User Manual, English Version, pp. 385-386, 2007

[Schaer et al., 2005]

Schaer, Thorsten; Jacobs, Jürgen; Scholl, Susanne; Kurby, Stephan: *DisKon-Laborversion Eines Flexiblen Modularen und Automatischen Dispositionsassistenzsystems.* In: ETR-Eisenbahntechnische Rundschau, Nr. 12, 2005, pp. 809-821

[Siefer, 2008]

Siefer, Thomas: Simulation. In: Ingo Arne Hansen, Jörn Pachl (eds.): Railway Timetable & Traffic Hamburg: Eurailpress, pp. 155-169, 2008

[Siefer and Radtke, 2006]

Siefer, Thomas; Radtke, A.: *Evaluation of Delay Propagation.* In: Proceedings of 7th World Congress on Railway Research, Montreal, 2006

[Tritschler et al., 2005]

Tritschler, Stefan; Cui, Yong; Dobeschinsky, Harry: *Störfallmanagement im ÖPNV*. In: Der Nahrverkehr, 2005, pp. 14-18

[UIC, 2016]

International Union of Railways (UIC): RailTopoModel – Railway Infrastructure Topological Model. In: International Railway Solution IRS 30100, Paris, 2016

[UIC, 2014]

International Union of Railways (UIC): RailTopoModel Railway Network Description – A Conceptual Model to Describe a Railway Network. 2014

[VIA-Con, 2016]

VIA Consulting and Development: *LUKS – Analysis of Lines and Junctions*, (16.12.2016). <http://www.via-con.de/en/development/luks>, Accessed 16.December 2016.

[Wende, 2003]

Wende, Dietrich: *Fahrdynamik des Schienenverkehrs*, Vieweg+Teubner Verlag, 2003

[Yuan, 2006]

Yuan, Jianxin: *Stochastic Modeling of Train Delays and Delay Propagation in Stations*, PhD thesis, TU Delft, 2006.